Imprint:

Copyright © 2017 GRIN Verlag
Print and binding: Books on Demand GmbH, Norderstedt Germany
ISBN: 9783668714861

This book at GRIN:

https://www.grin.com/document/420553

Simon Dedic

In vitro characterization of oncolytic virus in combination with STAT3 inhibition for improved therapy against pancreatic ductal adenocarcinoma

GRIN Verlag

Master thesis in Biology

Ludwig-Maximilians-Universität

In vitro characterization of oncolytic virus in combination with STAT3 inhibition for improved therapy agains pancreatic ductal adenocarcinoma

by

Simon Dedic

December 2017

Abstract

Being the most common malignancy of the pancreas, pancreatic ductal adenocarcinoma (PDAC) displays 5-year survival rates below 6% and ranks among the deadliest types of cancer, especially in the Western world. Several approaches have been explored to improve the outcome of this disease, but the highly complex biology and the aggressive nature of this malignancy require improved and novel therapeutic approaches.

Recently, oncolytic viruses have emerged as an innovative strategy to treat tumors. Exploiting defects in antiviral signaling in tumor cells, oncolytic viruses target and replicate only in cancer cells, while leaving healthy tissues generally unaffected. In particular, vesicular stomatitis virus has proven itself to be an effective oncolytic agent, due to its rapid life cycle and ability to rapidly lyse tumor cells.

In parallel, accumulating evidence has revealed the tumorigenic role of signal transducer and activator of transcription 3 (STAT3). Through its positive regulation of many oncogenic pathways responsible for cancer initiation, progression and metastasis, it has gained increasing interest as a novel therapeutic target.

Based on the outcome of previous studies, we hypothesized that a therapy or rVSV-GFP in combination with STAT3 inhibition in vitro would not only lead to an improved anti-tumor response in PDAC cells, but would also improve the safety in healthy tissues. Within the scope of this thesis, two-dimensional, three-dimensional as well as co-cultural in vitro studies were performed, with the aim of characterizing the effect of a combining these two therapeutic agents.Based on our results, we were able to show that both, rVSV-GFP and the STAT3 inhibitor S3I-201, have cytotoxic effects in PDAC cells in vitro. However, a combination therapy resulted in enhanced tumor cell killing in an additive manner, while attenuating viral replication in healthy primary pancreatic cells, being the case in. Further investigations in tumor spheroids revealed that these effects seem to be due to the induction of apoptosis, while the combination of both agents even improved the spread of apoptotic cells through the three-dimensional tissue architecture. In addition, we observed that the combination therapy significantly reduced the migratory and proliferative potential of both, PDACs and cancer-associated fibroblasts.

Substantiated by our results, we believe that combining oncolytic virus with STAT3 inhibitors is a novel treatment option for improving the dismal prospects of PDAC. This approach revealed not only cytotoxic effects on the cancer cells, but due to its anti-proliferative properties and its lytic effect on the stromal cells, it is represents a promising multi-mechanistic therapy for modulating the PDAC microenvironment and potentially resolving the tumor. This work provides preliminary data to support further testing in vivo and the potential future development for clinical translation.

Contents

I. Introduction

1.1 Pancreatic cancer

Pancreatic ductal adenocarcinoma, PDAC in short, is the most common malignancy of the pancreas. With a 5-year survival rate of less than 6%, it belongs to the deadliest types of cancer and is the fourth leading cause for cancer related deaths in the United States (Rossi, Rehman et al. 2014). Even though the most survival rates of cancer have been increased in the past few decades, PDAC still displays increasing incidence, especially in the Western world (Xu, Pothula et al. 2014). According to the latest statistics in epidemiology, PDAC is anticipated to become the second most common cause for cancer-related deaths right behind lung cancer in 2022.

The reason for such dismal prognoses is due to the fact that pancreatic ductal adenocarcinoma poses a challenge for both, clinicians and researchers. Already diagnosis of this disease is challenging, since PDAC exhibit symptoms only at very late stages (Kleeff, Korc et al. 2016). Besides owing to a lack of specific markers, there is no molecular approach to identify pancreatic cancer. Also medical treatment is limited as prevalent mutations are not really druggable yet. In addition, this disease exhibits an extremely aggressive nature, which results in rapid high level metastases (Sharma NK 2014). Furthermore, the tumoral environment is particularly relevant for PDAC, since its surrounding highly dense stroma supports cancer progression, while it impedes the accessibility of therapeutics and the immune system (Feig, Gopinathan et al. 2012).

Described above are just few explanations among many others, why PDAC is challenging not just on a molecular level but also in translational research and clinical application. To understand its extremely malignant and multifaceted nature, further insight into the evolution and expansion of pancreatic cancer is necessary.

1.1.1 PDAC development and progression

Besides risk factors as smoking, obesity and aging, also pancreatitis and genetic predisposition concur to the likelihood to develop pancreatic cancer (Klein 2012). Crucial for the development of PDAC are phenotypical as well as genotypical changes, both caused by particular mutations in the genome. The development of PDAC is a complex step by step process, of which each step is initiated by a specific mutation (Bardeesy and DePinho 2002). Once the first mutations appeared, genetic instability in these cells is accumulating, which drives the probability of the acquisition of new mutations in a positive feedback loop manner.

The first step in the emergence of this disease is the dedifferentiation of acinar cells to an embryonic progenitor phenotype in the pancreas. This initiating event is caused by epigenetic silencing of *Ptf1a*, an acinar cell identity factor, and the upregulation of *Pdx1*, responsible for pancreatic development and the maturation of β-cells (Pinho, Rooman et al. 2011).

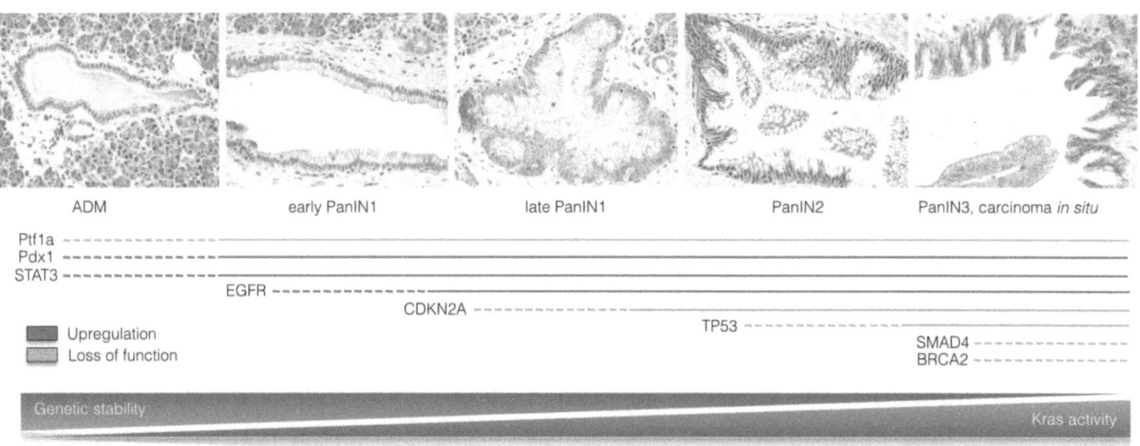

Figure 1: Different stages of PDAC development are dependent on particular mutations
Cancer development is initiated through acinar to ductal metaplasia, mainly induced by *Ptf1a* and *Pdx1* aberrations. *Kras* upregulation and EGFR overexpression initiates PanIN formation, which progression through its different states is dependent on further mutations in the genome (i.e. *CDKN2A*, *TP53*). Histological pictures were adapted from Brosens et al., 2015

Acinar cell dedifferentiation lays the foundation for the subsequent acinar to ductal metaplasia, ADM in short. If naturally occurring, for example in pancreatitis, this is a reversible mechanism which aids the regeneration of the cell after inflammation or an injury.

Besides Pdx1, SOX9 and Notch signaling are important factors for this transdifferentiation to a ductal-like cell (Storz 2017). But also inflammatory macrophages are capable of initiating this event by cytokine signaling. Hence, also IL-6 induces ADM through the activation of the JAK-STAT3 pathway (Yaqing Zhang and Zetter 2013). Furthermore, ADM can also be irreversible if mutations in *Kras* occur, which leads to the activation of oncogenic *Kras* and the initiation of pancreatic intraepithelial neoplasias (PanINs) (Shen, Wang et al. 2013). PanINs represent different progressive stages that need to be passed through for the development of PDAC, which are histologically characterized by morphological alterations compared to the healthy pancreatic duct. Subdivided into three different states, each of them are associated with specific genetic mutations.

As mentioned above, the first stage, PanIN1, is initiated by oncogenic mutations in *Kras* and its hyperactivation. This induces many pro-proliferative pathways which include but are not limited to the RAF-MEK-ERK and the PI3K-AKT pathway (Ferro and Falasca 2014). *Kras* activity is also accumulating by the overexpression of EGFR. With the progression of this stage, also the capability of senescence is lost through the loss of *CDKN2A*. The following loss of *TP53*, the most prominent and important tumor suppressor gene, is typical for PanIN2 and the first nuclear abnormalities. With the accumulation of genetic instability and the loss of two other tumor suppressor genes, *SMAD4* and *BRCA2*, the neoplastic progression reaches its last state PanIN3, also referred to as carcinoma *in situ* (Bardeesy and DePinho 2002). At this stage, the neoplasia is histologically hardly distinguishable from pancreatic ductal adenocarcinoma, which is then reached with further aberrations of the genome that enables the tissue of invasiveness and metastasis.

1.1.2 Microenvironment

Besides genetic mutations also PDACs microenvironment plays a crucial role not just for its development but its progression and maintenance as well. A mature pancreatic cancer is histologically defined as tumor islands that are surrounded by its dense stroma. Hence, the desmoplastic regions can account for up to 90% of this disease (Xie and Xie 2015). Being a hallmark of cancer, these regions are highly heterogenous. They consist of cellular components like cancer associated fibroblasts (CAFs), pancreatic stellate cells (PSCs) and immune cells, as well as acellular extracellular matrix (ECM), growth factors and cytokines. All these components form a complex network and the interplay with each other as well as the cancer cells is crucial for cancer initiation and progression (Xu, Zhou et al. 2016).

An important role in these interactions play cancer associated fibroblasts, which are known to promote the progression and metastasis of PDAC. This is achieved by the secretion of growth factors and inflammatory cytokines, the mediation of ECM formation and suppressing the immune system. Vice versa, CAFs are activated by the cancer cells. As a consequence, α-SMA, expressed by activated stromal cells, belongs to negative prognostic markers for PDAC in the clinics (von Ahrens, Bhagat et al. 2017).

As shown in Figure 1, *SMAD4* deficiency as well as *Kras* hyper activation play important roles in PDAC. This is also attributed to the interaction with its stroma. It was shown for this combinational mutation that it leads to an constitutive activation of the TGF-β, resulting in accelerated CAF and PSC proliferation, enhanced ECM production and eventually induced angiogenesis via the VEGF pathway (Ahmed, Bradshaw et al. 2017). Furthermore, the constitutive IL-6 production of the stromal cells over activates the STAT3 pathway in tumor cells in a paracrine manner. This pathway is commonly known for its role in oncogenesis as well as its capability of initiating metastasis via the induction of epithelial to mesenchymal transition (EMT) (Nagathihalli, Castellanos et al. 2016).

Besides many others, aberrational signaling via TGF-β and IL-6/STAT3 are examples to display the importance of the tumor-stroma-interactions within pancreatic ductal adenocarcinoma. As a result, many studies have taken this into account for the development of novel treatment options.

1.1.3 Current treatment options and their limitations

Both, treating PDAC as well as developing new therapeutic options display major challenges to clinicians and researchers. Surgery still remains the most effective option to cure pancreatic cancer completely. In former times, surgical resection was questionable due to high morbidity and mortality rates. Nowadays, however, increased experience led to improved surgery. As a consequence, successfully resected PDACs display 5-year survival rates of up to 20%, which is significantly higher but frankly still not promising (Kleeff, Korc et al. 2016). Additionally, pancreatic cancer barely exhibit symptoms at early stages and specific markers for this disease are non-existent. Given the aggressive nature, this results in advanced stages of PDAC and metastatic spreads throughout the whole body at the time of prognosis (Kamran, Patil et al. 2013). Therefore, surgery is rarely a possible treatment option when PDAC is diagnosed.

Another approach to counteract PDAC is chemotherapy. In the 1950s, 5-Fluorouracil (5-FU) was the mainline chemotherapeutic, resulting in mean survival rates of less than 6 months. This was improved with the finding of gemcitabine mono therapy that extended the survival rates to almost 1 year (Burris, Moore et al. 1997). The most noteworthy approach to optimize chemotherapy was FOLFIRINOX in 2011, a combinational approach consisting of 5-FU, irinotecan, leucovorin and oxaliplatin. Despite to the fact, that it displays significantly increased toxicity, its considerable increased survival benefit made FOLFIRINOX to one of the most widely used chemotherapeutic for advanced metastatic pancreatic cancer (Conroy, Desseigne et al. 2011). The problems with treating PDAC via chemotherapeutics is attributable to many different reasons. First, most of the prevalent mutations illustrated in Figure 1 are hardly druggable yet. This hampers the progress that is achieved by researchers, since the findings are difficult to apply translationally. Additionally, it was shown that PDAC is able to develop resistances to extant therapies, which again decreases the survival chances (Schober, Jesenofsky et al. 2014). Furthermore, PDACs microenvironment mentioned above is crucial for limiting the treatment efficacy. Its abundant stroma embed the tumor cells in a highly dense entity, limiting the vascular perfusion (Khan, Jaggi et al. 2015). As a result, adenomatous components of the tumor are reached in a marginally amount and chemotherapeutics are not able to access their target.

Owing to the many tumor supportive effects exerted, PDACs stroma became a commonly targeted treatment option for many researchers. However, due to contradictory findings it remains unclear if depleting the microenvironment of pancreatic cancer increases the survival. First approaches were conducted by inhibiting sonic hedgehog signaling (Shh), a protein responsible for the paracrine activation of stroma cells, which reduced desmoplasia and increased the survival rates in mice, but clinical trials were stopped because of their poor performance. Further studies for Shh ablation resulted in even more aggressive tumors with increased metastasis (Mei, Du et al. 2016). Other studies have tried to achieve this by the inhibition of matrix metalloproteinases (MMPs) (Zervox, Franz et al. 2000). MMPs are upregulated by cancer cells during EMT and are crucial for the initiation of metastasis. However, inhibiting MMPs resulted in contradictory and partly more aggressive tumors as well. To some extent this was also shown for targeting CAFs (Carr and Fernandez-Zapico 2016). Nevertheless, targeting stroma still eventuated in increased vascularity, which could potentially increase the efficacy of chemotherapeutics. Additionally, due to the tumor suppressive role of PDACs microenvironment, its ablation could possibly lead to an enhanced anti-tumoral activity of the immune system (Xie and Xie 2015). Recent studies have also suggested, that an approached equilibrium could be the solution, which would keep the stroma in rein but does not ablate it (Kota, Hancock et al. 2017).

Considering the many limitations that current treatment options display, the necessity of novel treatment options is obvious. Despite approaches like precision medicine and immunotherapy, oncolytic viruses could be one of those.

1.2 Oncolytic viruses

Oncolytic viruses (OVs) are either genetically modified or naturally occurring viruses that selectively target and replicate solely within tumor cells while not harming the healthy cells of the host (Altomonte and Ebert 2012).

The fact that particular viruses are able to kill cancer cells was recognized already in the early 20th century, when tumors regressed in patients after acquiring a natural infection. Afterwards, various approaches were undertaken to exploit the anti-tumoral effect of viruses. But even if clinical trials led to tumor regression, lack of knowledge in those days resulted mostly in failure of treatment (Hartkopf, Fehm et al. 2012). Nowadays, knowledge increased in immunology and tumor biology, as well as we have the possibility of genetically engineering the genome. Hence, this therapeutic approach gained much interest, especially in the last decade as two genetically engineered OVs have been approved for cancer therapy. In 2005, Oncorine, or H101, an adenoviruses with a deletion in the E1B-gene, was approved in China for the treatment of esophagus and head and neck cancer (Yu and Fang 2007). More recently, T-Vec, a herpes simplex virus that was genetically armed with granulocyte macrophage colony-stimulating factor (GM-CSF), was approved by the FDA in the USA in 2015 and one year later in Europe and Australia for treating melanoma (Kohlhapp, Zloza et al. 2015).

Even though further investigations have to be done for the complete comprehension, OVs are thought to kill cancer cells via two steps.
First, they replicate selectively within tumor cells as the have many traits that provide a selective advantage for the virus. Besides hallmarks of cancer like sustained cell proliferation, resisting cell death and avoiding the immune system, the most important abnormality is that tumor cells are defective in their protection mechanisms against viral infection (Kaufman, Kohlhapp et al. 2015). For example, the Interferon (IFN) signaling pathway, which is the very first antiviral response as soon as cells get infected. Additionally, tumor cells can be resistant to signaling of the protein kinase R (PKR), responsible for the detecting and responding to the viral genome (Singh, Doley et al. 2012). Although OVs are also able to infect healthy tissues, they get cleared by the antiviral response, while retaining and further replicating in the neoplastic cells.
As a second step, tumor cells get lysed and release tumor antigens and OVs again. While the virus is then able to distribute further throughout the tumor and infect the cells in a positive feedback loop manner, the released tumor antigen potentially activates a host immune response against the cancer (Jhawar, Thandoni et al. 2017).

OVs exhibit advantages which are missing in many current therapeutic options. They are highly selective for tumor cells while just minimal toxicity was detected for some viruses (Chiocca 2002). Additionally, as the pharmacokinetics of classical therapeutic drugs rather decrease over time, OVs are able to amplify themselves during treatment, resulting in an enhanced anti-tumoral response. Besides, today's advances in genetic engineering allows the modification of these therapeutic agents, enabling to improve the tumor killing effect and reducing toxicity in healthy tissues at the same time (Goldufsky, Sivendran et al. 2013). Particularly remarkable is the fact, that OV treatment could eventually counteract not just tumors but also metastasis through both active infection of them as well as priming the immune system against tumor antigens (Park and Choi 2016). Lastly, since OVs exploit many different oncogenic features, development of resistances are rather unlikely and weren't seen so far.

Contrary to the many positive effect of OVs, they also show some disadvantages. Especially delivery of the virus hampers therapeutic efficiency. Intravenous or intraarterial administration of the virus mostly leads to its rapid recognition and destruction via the complement system as well as the innate immune system. Intratumoral injection is hence usually the best option for delivery, but would make the treatment conditional on the tumor site, which is not always accessible for injections (Seymour and Fisher 2016). Furthermore, the tumor microenvironment, which plays a crucial role especially for pancreatic cancer as mentioned above, is another hurdle by limiting OVs ability to spread throughout the whole tumor (Wojton and Kaur 2010). And as these therapeutic agents are indeed quite harmless for healthy tissues, there were still some cases were toxicity was detected, which is still the reason for the failure of some clinical trials.

Nevertheless, OVs belong to the major promising approaches and are considered as breakthrough in immunoncology (Fukuhara, Ino et al. 2016). Within the scope of this thesis, we were working with the Vesicular stomatitis virus (VSV).

1.2.1 Vesicular stomatitis virus

VSV is a bullet shaped virus that belongs to the family of the Rhabdoviridae. Its genome consists of 11 kb single stranded RNA and as depicted in Figure 2, it is encoding for 5 different proteins: the Glyco- (G), Matrix- (M), Large- (L), Phospho- (P) and Nucleoprotein (N).

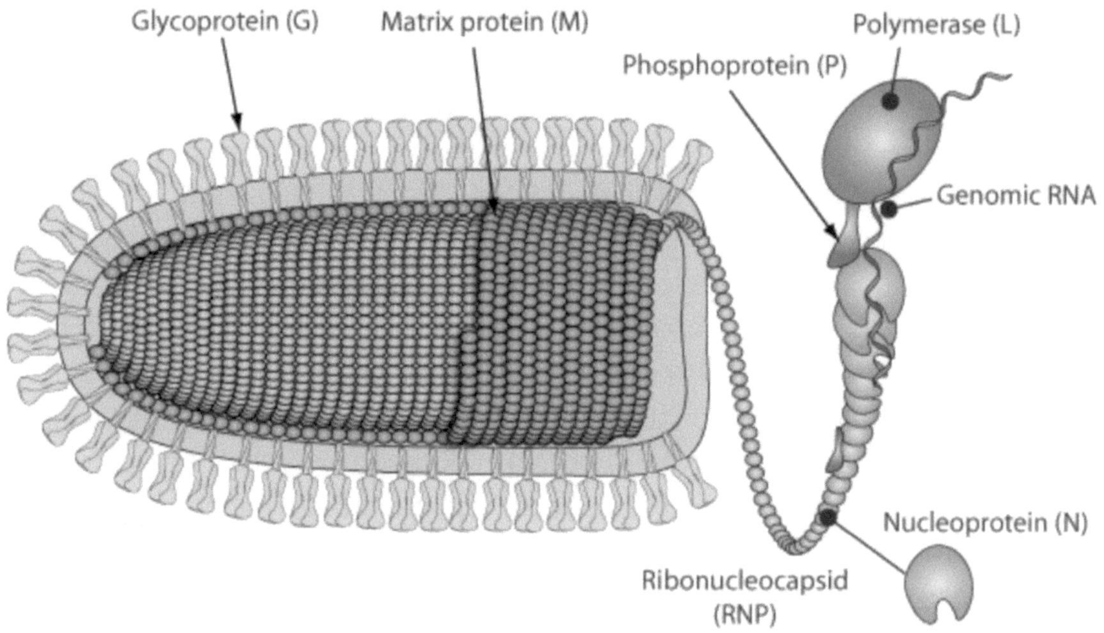

Figure 2: Architecture of Vesicular stomatitis virus
The bullet shaped envelope of VSV is covered with Glycoproteins, responsible for the attachment to the host cell. Within the virion, the genomic RNA is tightly encapsidated by the Nucleoproteins. The phosphoprotein together with the Polymerase (encoded by the L-protein) is translating the genome into the mRNA. The Matrixprotein serves the blocking of the hosts cellular response. Picture was adapted from Viralzone (viralzone.expasy.org)

The G-protein allows VSV to entry the host cell via binding LDL receptors on its membrane, which is then uptaken through endocytosis. The N-protein tightly encapsidates the genome of VSV, which is recognized solely by the viral RNA polymerase, encoded by the L-protein, which translates the mRNA together with the P-protein. The purpose of the M-protein is to inhibit cellular responses of the host cell to the infection. This is accomplished by migrating to the nuclear envelope and interacting with Rae1 and Nup98, factors that are necessary for the nucleocytoplasmic trafficking of cellular mRNAs and snRNAs. Natural hosts of VSV include mostly horeses, cattle and pigs. VSV infection in humans are rarely linked with the occurrence of symptoms and are limited to people with direct and frequent contact with this virus (Heiber, Xu et al. 2011).

Contrary to many other OVs, VSV exhibit many natural occurring advantages. To begin with, VSV is a virus model that was already extensively investigated, which alleviates to conceive coherences more quickly. Additionally, it infects the cells mostly independent of their cell cycle, which again facilitates the work with this virus. Since it is a single stranded RNA virus and it replicated within the cytoplasm, there is no risk that the viral genome integrates into the hosts DNA. Another aspect, which is important when trying to establish a virus for oncolytic therapy, is the lack of pre-existing immunity in humans, otherwise it would be immediately recognized by the immune system and eventually hamper the delivery of OVs. Lastly and more importantly, VSVs small genome is easy manipulated, allowing the insertion of oncoselective and oncotoxic factors, as well as the deletion of genes that could potentially harm healthy tissues (Hastie and Grdzelishvili 2012).

Many different approaches have been done to improve VSV. Obviously, a VSV-GFP variant was engineered to facilitated the visibility of VSV under the microscope. Furthermore, neurotoxicity was shown in rodents and also once in humans, which would be a dramatic side effect in the clinic. To counteract this, a VSV-ΔM51 variant was genetically modified, resulting in increased safety (Simovic, Walsh et al. 2015). Among many other, there are also variants modulating the immune response, encoding suicide genes or cancer suppressor genes (Fernandez, Porosnicu et al. 2002). Another approach is to combine VSV with drugs that potentially increase its killing effect against the tumor while increasing the viral safety at the same time.

1.3 Signal transducer and activator of transcription 3

The signal transducer and activator of transcription 3 is one of the seven members that belong to the STAT family. As the name already implies, the purpose of these proteins is to transduce a signal from the cell surface to the nucleus, where transcription of various genes becomes activated.

Encoded by chromosome 12, the STAT3 protein consists of seven structural and functional distinct domains. The first one is the NH2-domain that forms homotypic dimers as long as STAT is not phosphorylated and in an inactive state. Directly adjacent to it is the coiled-coil domain, interacting with regulatory proteins. Furthermore, there is the DNA-binding domain, responsible for the localization on the promoter on the genome, as well as the SH2-domain, mediating dimerization when activated. These two domains are linked by the linker domain, hence leading to a proper conformation of the protein. Crucial for the activation of the STAT3 protein is the tyrosine activation domain, containing a conserved tyrosine residue which can be phosphorylated at position 705. Lastly, the activation of downstream pathways is then induced by the transactivation domain through the interaction with transcriptional regulators (Darnell 1997).

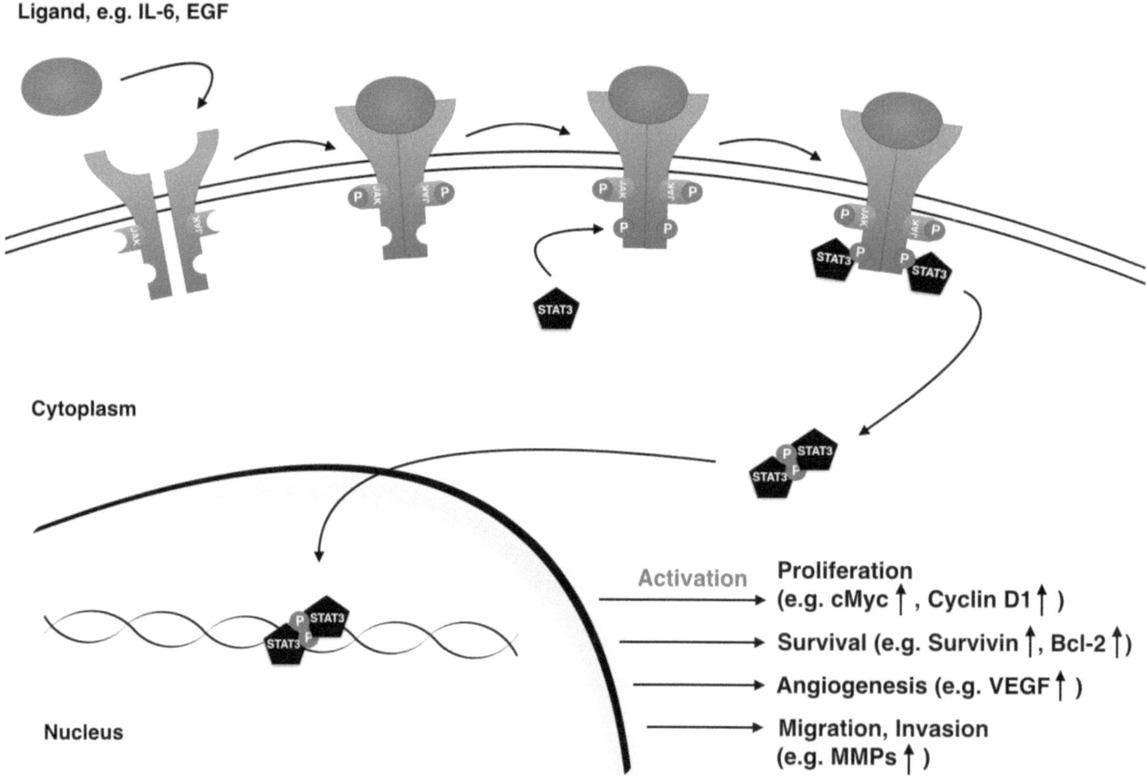

Figure 3: Activation of STAT3 results in transcription of oncogenic proteins
Once the appropriate receptors are activated by various ligands, a phosphorylation cascade leads to the activation of STAT3. The phosphorylated STAT3 then dimerizes, translocated to the nucleus and acts as transcriptional factor. In that way, many oncogenic genes become activated, resulting in e.g. enhanced proliferation, survival, angiogenesis and invasive potential.

Figure 3 illustrates the mechanism of action by which STAT3 becomes activated. This can be induced by various cytokines and growth factors like IL-6 or EGF. The receptors responding to such ligands are usually tyrosine kinases, often associated with Janus activated kinases (JAKs). After binding of the ligand, its receptors dimerize and lead to the phosphorylation of JAKs. Subsequently, adjacent tyrosines on the receptor become phosphorylated, which are then bound by the SH2-domain of STAT3 proteins. Once two STAT3 monomers are activated by the phosphorylation at tyrosine 705 (Y705), dimerization and translocation into the nucleus follows. STAT3 acts as transcription factor on the DNA and induces transcription of many downstream effectors that possess oncogenic potential (Yue and Turkson 2009).

STAT3's tumorigenic role is multifaceted and occurs at all different levels of cancerogenesis, beginning from the induction of cancer, over its progression, up to metastasis. Besides its role in malignant cell transformation, STAT3 is crucial for increased proliferation through upregulation of *cMyc* and *Cyclin D1* (Xiong, Yang et al. 2014). Additionally, it prevents apoptosis by inducing the transcription of *Bcl-2* plus negatively regulating p53 (Niu, Wright et al. 2005). Furthermore, STAT3 is essential for migration, invasion, intravasation and angiogenesis, mechanisms that are all determining for metastasizing. Nevertheless, especially remarkable for pancreatic cancer is its function for the paracrine communication with the tumor microenvironment (Nagathihalli, Castellanos et al. 2016). It was shown that STAT3 activates fibroblasts and enhances their proliferation through the TGFβ pathway, resulting in fibrotic reactions. At the same time, CAFs secrete high amounts of IL-6, one of the most important cytokine for the activation of STAT3, which results in a positive feedback loop (Zhu, Zhang et al. 2014).

Although inactivating STAT3 leads to embryonic lethality in mice, it was shown that its role in healthy adult tissue is unimportant (Kamran, Patil et al. 2013). Hence, given the high oncogenic potential of STAT3, its inhibition gained increasing interest for translational application in the clinic. As a consequence, the inhibitory capability of both naturally occurring substances and synthetically derived chemical agents were investigated. Within the scope of this thesis, we were working with the chemical probe inhibitor S3I-201, also called NSC 74859.

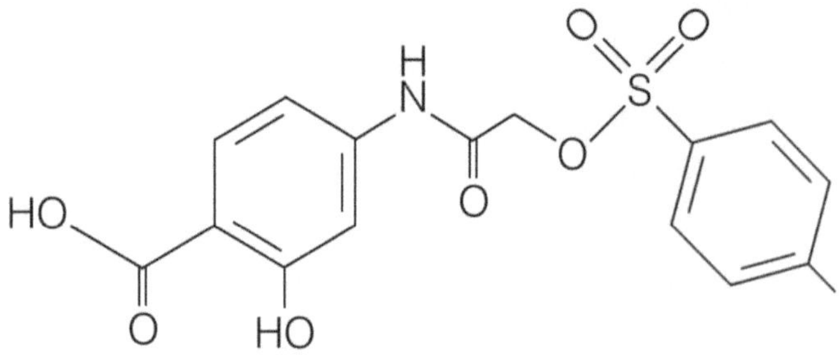

Figure 4: Chemical structure of S3I-201

S3I-201, or NSC 74859, prevents dimerization of activated STAT3 monomers by docking to the SH2-domain

The inhibitor depicted in Figure 4 above targets the critical step which leads to the dimerization complex of two activated STAT3 proteins. By docking directly to the SH2-domain, nuclear translocation and hence, transcriptional activation can not happen anymore. It was already shown for S3I-201 that it acts antiproliferative and proapoptotic on malignants cells while being harmless to healthy cells. Furthermore, it was even shown that it can effect tumor regression in xenograft models (Siddiquee, Zhang et al. 2007).

1.4 Aim of this thesis

As mentioned before, because of the poor survival rates and the aggressive nature of PDAC, this disease is considered as a very severe diagnosis. Further investigations gave us deeper insights into this disease, showing us the high complexity of the mutational cascades in PDAC initiation and progression (Kleeff, Korc et al. 2016). Also the importance of the tumor microenvironment was clarified, but as different studies led to paradoxic results in targeting the stroma, it gives us evidence that it is just poorly understood yet (Carr and Fernandez-Zapico 2016). Given the limitations of current treatments and the failure of various novel approaches, a necessity of further investigating new therapeutic options is unquestionable.

With the approval of T-Vec and Oncorine, oncolytic viruses gained increasing interest as novel treatments in oncology. Especially the many naturally occurring advantages and the deadly effect of Vesicular stomatitis virus against a variety of cancer types promises strong antitumoral effects (Hastie and Grdzelishvili 2012).

Additionally, STAT3 inhibitors became more encouraging as evidences accumulate that STAT3 plays a crucial role at different levels of carcinogenesis. Particularly its function in metastasis and its cross talk with the tumor microenvironment makes targeting STAT3 attractive to keep advanced tumors in rein (Nagathihalli, Castellanos et al. 2016).

A previous study of our lab examined the effect of VSV together with the STAT3 inhibitor S3I-201 as a combinational therapy against hepatocellular carcinoma. It was shown for this deductive approach that it reduces the toxicity of VSV in both healthy primary cells as well as glia cells *in vitro*. At the same time, viability assays showed even enhanced antitumoral effects that VSV has with increasing concentrations of S3I-201. Taken together, this study suggested that combining VSV with a STAT3 inhibitor, would not just increase the tumor killing effect, but would also increase the safety of this oncolytic virus, as this is the most limiting factor for translational application (Marozin, Altomonte et al. 2015).

Since there is accumulating evidence that STAT3 is constitutively activated in PDAC and important for the progression and maintenance of this disease, the goal of this thesis was to establish a PDAC model in the lab to further investigate and characterize if the synergistic effect of VSV with STAT3 inhibition would also be effective on this cancer type. Aiming to unravel as many characteristics as *in vitro* studies allow, this study utilized not just 2D and 3D systems, but also cocultural approaches on both molecular and morphological levels.

II. Material

2.1 Cell lines

Abbreviation	Type	Host
110299	Pancreatic tumor cells	Pdx1-Cre; Kras^{G12D} Mouse
530202	Pancreatic tumor cells	Pdx1-Cre; Kras^{G12D} Mouse
14128	Pancreatic tumor cells	Pdx1-Cre; Kras^{G12D} Mouse
F3 338	Cancer associated fibroblasts	Sm22CreERT; Prrx1; tdEG Mouse
F3 704	Cancer associated fibroblasts	Sm22CreERT; Prrx1; tdEG Mouse
PDC	Pancreatic ductal cells and duct-like cells	isolated out of a healthy mouse pancreas
BHK-21	Baby hamster kidney cells	Hamster

2.2 Reagents

Reagent	Company
1-Step Transfer Buffer	Thermo Scientific
10 x Cell Lysis Buffer	Cell Signaling Technology
10 x Tris/Glycine Buffer	BioRad
10 x Tris/Glycine/SDS Buffer	BioRad
10% Tween 20 Solution	BioRad
2 x Laemmli Sample Buffer	BioRad
3,3,5 - Triiodo - L - Thyronine	Sigma-Aldrich
30% Acrylamide/Bis Solution	BioRad
APS	Sigma-Aldrich
Blotting - Grade Blocker	BioRad
Bovine Pituitary Extract	Corning
CellTiter 96 AQ$_{ueous}$ MTS Reagent	Promega
CellTiter Glo 3D Reagent	Promega
Cholera Toxin	Sigma-Aldrich
Collagen Type I Rat Tail	Corning
Collagenase Type IV	Worthington - Biochem

Complete Protease Inhibitor Cocktail	Roche
D - Glucose	Sigma-Aldrich
Dexamethasone	Sigma-Aldrich
DMEM F12	biowest
DMEM High Glucose	biowest
DMEM Low Glucose	biowest
DPBS	PAN Biotech
Epidermal Growth Factor	Corning
FBS Superior	Merck
G-MEM BHK-21	gibco
ITS+ Premix	BD Biosciences
L - Glutamine	PAN Biotech
Nicotinamide	Sigma-Aldrich
Nu-Serum IV	Corning
OptiPro Serum Free Medium	gibco
Penicillin - Streptomycin	PAN Biotech
Phenazine Methosulfate	Sigma-Aldrich
PhosSTOP Phosphatase Inhibitor Cocktail	Roche
Precision Plus Protein Dual Color Ladder	BioRad
Resolving Gel Buffer	BioRad
Restore PLUS Western Blot Stripping Buffer	Thermo Scientific
SDS Solution 10%	BioRad
Soybean Trypsin Inhibitor Type I	Sigma-Aldrich
Stacking Gel Buffer	BioRad
TEMED	BioRad
Trypsin / EDTA	PAN Biotech
β - Mercaptoethanol	BioRad
Passive Lysis Buffer 5x	Promega
Amersham ECL Western Blotting Reagent	GE Healthcare
Lipofectamine LTX and Plus Reagent	invitrogen

2.3 Antibodies

Specificity	Host	Company
pSTAT3 Y705 XP(R)	Rabbit	Cell Signaling Technology
Actin	Mouse	Sigma-Aldrich
7-AAD	Mouse	Novus

2.4 Consumables

Consumable	Company
Spheroid Microplate Ultra Low Attachment 96 well	Corning
Transwell Permeable Supports, 0.4 µM membrane	Corning
96 well plate	Sigma-Aldrich
24 well plate	TPP
6 well plate	TPP
75 cm² Cell Culture Flask	Sigma-Aldrich
LabTek 4 well Chamber Slide System	nunc
CL-Xposure Film 18 x 24 cm	Thermo Scientific
Plasmid Purification Kit	QIAGEN

2.5 Appliances

Appliance	Company
Pierce Fast Semi-Dry Blotter	Thermo Scientific
X-ray Film Processor	OPTIMAX
gallios FACS	Beckman Coulter
Nanodrop Lite Spectrophotometer	Thermo Scientific
PowerPac Basic	BioRad
PowerPac HC	BioRad
Confocal Microscope SP8	Leica
Axiovert 40 CFL Fluorescence microscope	Zeiss
ELISA reader sunrise	Tecan
JuliBr Live Cell Movie Analyzer	Nanoentek
Glomax Luminometer	Turner Biosystems

III. Methods

3.1 Cell culture

All cultures were maintained at 37°C, 90% humidity and 5% CO2.

Tumor cell lines were derived from Pdx1-Cre; KrasG12D mice, a genetically engineered mouse model, which is characterized by a slow progression of PanIN lesions until they reach the stage of invasive and metastatic panreatic ductal adenocarcioma. The cell lines used, 110299, 530202 and 14128, represent 3 individual cell clones that were isolated from primary PanIN lesions from these mice. The PDAC cell lines were cultured in Dulbecco's Modified Eagle Medium High Glucose (DMEM), additionally supplemented with 10% Fetal Bovine Serum, 100 U/ml Penicillin and 100 µg/ml Streptomycin.

Cancer associated fibroblasts were isolated from Sm22-CreERT2, Prrx1, tdEG mice, a transgenic mouse model in which the tamoxifen inducible Cre-recombinase is under the control of the transgelin, or smooth muscle 22 promoter. Within the scope of this thesis we worked with two CAF cell lines, F3 338 and F3 704, originated from two different mice. CAFs were cultured in medium consisting of 40% Dulbecco's Modified Eagle Medium Low Glucose, 40% Dulbecco's Modified Eagle Medium F12, 20% Fetal Bovine Serum plus 100 U/ml Penicillin and 100 µg/ml Streptomycin.

Primary pancreatic duct cells (PPC) and ductal-like cells were isolated from the pancreas of healthy mice, as previously described (Reichert, Takano et al. 2013). Prior to seeding, dishes were precoated with a collagen solution, prepared as described in Table 1 below. After the precoating, covered plates were incubated for at least 3 hours at 37°C to allow the collagen solution to solidify. Collagen coated dishes were always used immediately after solidification.

Reagent	Concentration	Volume for 10 ml
Corning Collagen Type I	3.59 mg / ml	6.5 ml
DPBS	10 x	1 ml
sterile dH_2O	1 x	2.335 ml
NaOH	1 N	0.165 ml

Table 1: Composition of the collagen solution
PDC cells required culture systems that were precoated with collagen, comprised of Collagen Type I, DPBS, sterile dH_2O and NaOH

PDC were cultured in a special composition medium, described in Table 2.

Reagent	Final concentration
3,3,5-Triiodo-L-Thyronine	5 nM
Bovine Pituitary Extract	25 µg / ml
Cholera Toxin	100 ng / ml
D-Glucose	4.7 mg / ml
Dexamethasone	1 µM
DMEM F12 medium	1 x
Epidermal Growth Factor	20 ng / ml
ITS+ Premix	1 x
Nicotinamide	1.22 mg / ml
Nu-Serum IV	5 %
Penicillin	100 U / ml
Soybean Trypsin Inhibitor	0.1 mg / ml
Streptomycin	100 µg / ml

Table 2: Composition of PDC full medium
Culturing of primary ductal cells required many additional reagents. Formula for this full medium was adapted from Reichert et al.

In order to harvest the cells, the supernatant was aspirated, and PPC containing collagen was scraped off the dish and put into a 50 ml tube, containing DMEM F12 with 1.5 mg / ml Collagenase Type V. For the purpose of dissolving the gel, these tubes were rotated for 15 min at 37°C. Afterwards, cells were centrifuged at 500 rcf for 5 min. The resulting pellet was trypsinized for 5 min and then resuspended in PDC full medium.

3.1.1 3D culture systems

Another cell culture technique utilized in this work was the establishment of 3D cell culture systems. Two different approaches were applied to form tumor spheroids, depending on the experimental endpoint.

The first method utilized Ultra-Low Attachment (ULA) 96-well plates, in which the coating inhibits the attachment of cells, forcing them into suspension. As a consequence, cells adhere to each other, forming highly reproducible, homogenous and round spheroids.
Cells were counted and diluted to the desired concentration. Cell suspensions were transferred to the ULA plates, and spheroids were cultured at 37°C and 5% CO2 for at least 3 days. Culturing for a longer time period was possible, but

required renewal of the medium. The number of seeded cells was quite adjustable. However, care has to be taken that cell number positively correlates with spheroid size and cohesiveness, hence also influencing the virus accessibility. This approach was used for viability assays.

Another method for the formation of a 3D system was achieved by culturing the cells in Matrigel. This is a special protein mixture resembling the compounds of the complex extracellular environment of tissue. In addition, its gelatinous consistency forces the cells to form 3D complexes. Since this not only enables spheroid formation, but the culturing of theses cells in 3D for a longer time period, this spheroids produced in this context are often referred to as organoids. This approach was used for investigations utilizing confocal microscopy.

Frozen Matrigel was thawed on ice. After counting, cells were adjusted to an appropriate cell density of 5000 cells / ml, followed by centrifugation at 500 rcf and for 5 min. Supernatant was discarded, the cell pellet was resuspended in 50 µl Matrigel, which was immediately transferred to a chamber slide. Matrigel drops were solidified at 37 °C for at least 30 min prior to overlay with medium. Cell culture medium was renewed every second day. To allow the cells to adjust to the environmental conditions and to form proper three-dimensional complexes, treatment was applied after a minimum of 5-7 days in culture.

3.1.2 Co-Culture Systems

Co-culturing of cells was achieved by two different approaches.

The first technique was a rather indirect technique, based on culturing tumor cells in the conditioned media (CM) of fibroblasts. CM was obtained from fibroblast cultures by harvesting the supernatant after two days of growth. To remove all residual cellular components, CM was filtered through filter membranes with a pore size of 0.4µm. CM was either used directly or was stored at -80°C.

For the experimental setup, tumor cells were cultured for at least two days in the CM of the fibroblasts, to ensure proper adjustment of the cells to the changing conditions. After this, viability assays were conducted, with tumor cells cultured in the medium conditioned by tumor cells serving as a control.

Another method, which ensured the direct co-culture of tumor cells with fibroblasts, was conducted using the Transwell Permeable Supports from Corning. These 6-well plates contain inserts with thin membranes containing 0.4 µm pores that allow the exchange of molecules between the cells cultured in the upper chamber and the lower chamber, while providing a physical boundary between the two cell types. To improve attachment of the cells, inserts were equilibrated for at least 1 hour with medium in the incubator. After cells were seeded, treatment followed as soon as 70-80% confluency was reached. For viability assays, inserts were moved into a fresh 6 well plate. In that way, MTS solution could be added to each cell line separately. Supernatant was transferred to a 96 well plate and measured using a colorimetric plate reader.

3.2 Viability assay

In order to reach 70-80% confluency at the time of treatment, cells were seeded in an appropriate amount. Post-treatment, cell viability was determined via the MTS assay.

CellTiter 96 AQueous MTS reagent powder was dissolved in PBS and adjusted to its optimum pH of 6.0 - 6-5. Before usage, MTS solution was mixed with a PMS solution in a ratio of 1:20.

This combined MTS/PMS solution was added to the cell culture medium of the samples in a ratio of 1:5. After incubation for 2-3 hours at 37°C and 5% CO_2, the absorbance at 490 nm was recorded via an ELISA reader.

For the quantification of the viability, untreated controls for normalization and blank controls were taken into account.

3.3 Tissue Culture Infection Dose 50 (TCID$_{50}$)

TCID$_{50}$ is a procedure for quantifying the infectious viral titers that can cause cytopathic effects (CPE) in tissue culture.

Baby hamster kidney (BHK-21) cells were seeded on a 96-well plate and cultivated until 70-80% confluency was reached. Afterwards, viral samples were serially diluted in 10-fold and applied in quadruplicate to the BHK-21, followed by incubation for 2-3 days at 37°C and 5% CO_2. Subsequently, viral titers were calculated by microscopic analysis to determine the last dilution at which CPE could be observed in all four wells and the percentage of wells with CPE at the following dilution. Viral titers were calculated using a statistical Excel spreadsheet (see Appendix).

3.4 Western blotting

Samples for Western blotting were harvested and centrifuged at 500 rcf for 5 min. The resulting pellet was resuspended in 1 x Cell lysis buffer supplemented with phosphatase inhibitor and Complete Protease Inhibitor Buffer. Lysates were stored at -20°C.

To prepare the samples, protein concentrations were recorded via a Nanodrop Spectrophotometer and equivalent amounts of protein were diluted 1:1 with 2 x Laemmli Sample Buffer containing β - Mercaptoethanol. For denaturing, samples were boiled at 95°C for 10 min. After centrifuging at 14.000 rcf for 5 min, samples were ready for loading on the gel.

Gels were self-poured based on the BioRad system. The composition of the gels is described in Table 3.

Reagent	7.5% Running Gel	10% Running Gel	Stacking Gel
30% Acrylamide	5 ml	6.6 ml	660 µl
Resolving Gel Buffer	5 ml	5 ml	-
Stacking Gel Buffer	-	-	1.26 ml
SDS	200 µl	200 µl	50 µl
dH20	9.7 ml	8 ml	3 ml
TEMED	10 µl	10 µl	5 µl
10% APS	100 µl	100 µl	25 µl

Table 3: Formula for self-poured gels
The formula is based on the pouring of two individual gels. Within the scope of this thesis, gels containing either 7.5% or 10% Polyacrylamide were used.

After gels fully polymerized, an amount of 50 µg protein in total was loaded from each sample to the wells and run at 100 V for approximately 90 min. Subsequently, the gel was sandwiched with a nitrocellulose membrane between two Whatman filter papers in order to transfer the proteins. This was performed either via the semi-dry or the wet method. Semi-dry transfer was conducted at 25V for 10 min via the Pierce Fast Semi-Dry Blotter after equilibrating all of the components in 1-Step Transfer Buffer. The wet transfer was performed in a tank at 100V for 60 min with Tris/Glycine Transfer Buffer that was supplemented with methanol. As a next step, the membranes with transferred proteins were rotated for at least 30 min in PBST + 5% milk to block unspecific binding. This was followed by incubation in primary antibody, which was diluted in PBST, at 4°C over night. On the next day, free primary antibodies were washed away 3 x 5 min in PBST. Next, incubation in secondary antibody, which was conjugated to a horseradish peroxidase, took place for at least 1 h at RT. In order to visualize the bound antibodies, membranes were incubated with ECL Western Blotting Reagent for 1 min. They were subsequently transferred to a Hypercassette and exposed to films in a dark room. Signaling strength, and hence exposure time, was variable for each protein and dependent on the amount of protein. In case of reutilizing the membrane again for the detection of further proteins, it was first washed with PBS and then stripped for a maximum of 5 min, using the Restore PLUS Western Blot Stripping Buffer, before it was blocked again and incubated with the next primary antibody.

3.5 Flow cytometric analysis

Cells were grown in 6-well plates and treated with STAT3 inhibitor or VSV-GFP as soon as 70-80% confluency was reached. Afterwards, cells were trypsinized and centrifuged at 500 rcf for 5 min. The cell pellet was resuspended in 4% PFA and incubated for 10 min at 4°C in order to assure fixation of the samples. Following a second centrifugation step, cells were resuspended in FACS buffer (1% FBS in PBS) and transferred into FACS tubes. For staining the cells, antibody of desire was diluted 1:100 in the FACS buffer and incubated for 10 min at 4°C. Subsequently, washing in PBS followed. After centrifuging the cells down, they were again resuspended in FACS buffer.
Samples were analyzed via the Gallios Flow Cytometer. Subsequent quantification was performed with the aid of the FlowJo software.

3.6 Confocal microscopy

Confocal microscopy served the purpose of analyzing the 3D cultures grown in Matrigel. After treating the cells for the desired period of time, medium was aspirated and cells were washed with PBS. Subsequently, cells were fixed by rocking in 4% PFA for 30 min at RT. After discarding the PFA, another three washing steps followed. In order to permeabilize the spheroids, they were incubated for 30 min at RT in permeabilization solution. Desired primary antibodies were diluted 1:1000 in permeabilization solution and incubated on the cells over night. On the next day, antibody solutions were aspirated and cells were washed 3 x 10 min in permeabilization solution at RT. Secondary antibody antibodies was were diluted 1:400, and DAPI was diluted 1:1000 and incubated on the cells again over night. On the third day, the antibody was aspirated again and three washing steps followed. Finally, the chamber cassette surrounding the samples was gently removed and one drop of fluorescent mounting medium was placed onto the slide, following by putting a coverslip on top. As soon as the mounting medium spread out on the whole slide, borders were sealed with nail polish. Slides were imaged upside down using the inverted laser-scanning confocal microscope SP8 from Leica.

3.7 Luciferase assay

3.7.1 Plasmid DNA preparation

50 µl of DH5α competent E.coli was thawed on ice, before 1 µg of plasmid DNA was added to them. After gently mixing, the cells were incubated for 30 min on ice. Afterwards, a heat shock for 20 seconds at 42°C was performed, followed by a 2 min incubation on ice. Subsequently, 950 µl of pre-warmed LB medium was

added, and cells were incubated at 37°C for at least 1 hour on a heat block with shaking at 225 rpm. 50 μl of this cell suspension was spread on an LB agar plate, containing Ampicillin, and incubated at 37°C over night.

On the next day, one colony was picked and grown in LB containing Ampicillin for at least 2 h at 37°C in a bacterial shaker, until it was transferred into 150 ml LB plus Ampicillin and incubated at 37°C over night. Extraction of the plasmids was conducted via the Plasmid Purification Kit from Qiagen. DNA concentrations were measured with the Nanodrop Spectrophotometer.

3.7.2 Transfection

Tumor cells were plated on a 24-well plate at a cell number of 1x105 cells/well, in order to reach 70-80% confluency at the transfection time point. 2 μl Lipofectamine LTX agent was diluted in 50 μl OptiMEM medium, and 500 ng DNA and 5 μl Plus Reagent were diluted separately in 50 μl OptiMEM medium. The diluted DNA was then added to the diluted Lipofectamine and incubated for 5 min at RT, before they were added to the cell containing wells. After overnight incubation, the medium was exchanged.

3.7.3 Interferon induction and response

Prior to transfection, tumor cells were left untreated or were pretreated with 100 μM STAT3 inhibitor. Afterwards, cells were transfected with plasmids containing a firefly luciferase gene under the control of either an Interferon Sensitive Response Element (ISRE) promoter or an IFN-β promoter. Both plasmids were cotransfected with a plasmid expressing the Renilla luciferase to serve as a normalization control for transfection efficiency. Subsequently, ISRE promoter-transfected cells were left unstimulated or treated with MOI 0.01 VSV-ΔM51 or 1000 U universal Interferon type I, while IFN-β promoter- transfected cells were left untreated or stimulated with MOI 0.01 VSV-ΔM51 and 1 μg polyI:C. After a second overnight incubation at 37°C, cells were lysed in 1 x Passive Lysis Buffer. Promoter activity was measured via the Dual Glo Luciferase Kit from Promega.

3.8 Scratch assay

Cells were grown in 6-well plates until they reached a confluency of 70-80%. After treating the cells, a scratch in the confluent monolayer was made, using a 200 μl pipette tip. The region with this resulting scratch was imaged via the JuliBr Live Cell Movie Analyzer at 37°C and 5% CO2 over the course of 72h. The image stack was later assembled to a film using iMovie.

IV. Results

4.1 PDAC cells are susceptible to rVSV-GFP-mediated oncolysis *in vitro*

Previous findings in our lab revealed an anti-tumoral effect of VSV in tumor cells. However, these experiments were conducted in hepatocellular carcinoma (HCC) cells. Aiming to establish a pancreatic model in the lab, the first step was to determine the effect of VSV on PDAC cells.

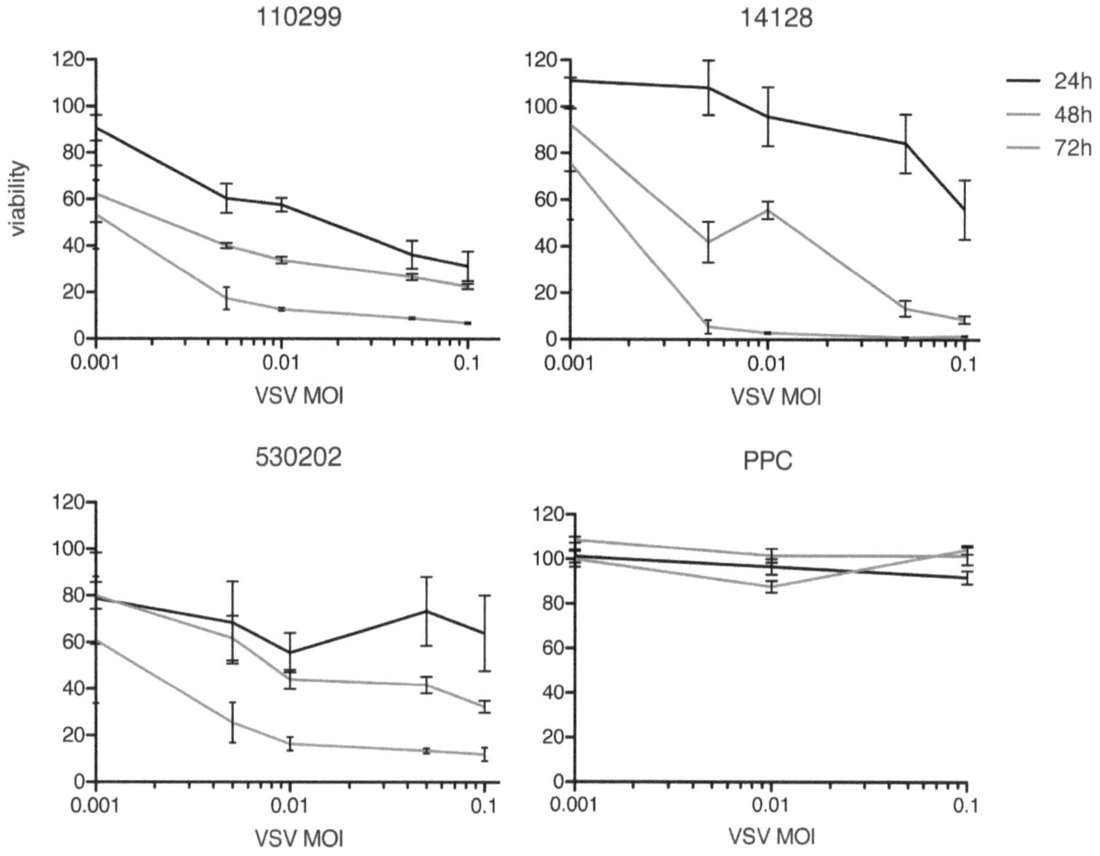

Figure 5: VSV kills PDAC cells but not PPCs
Three different PDAC cells lines, as well as PPCs, were infected with VSV at increasing MOIs (0.001, 0.005, 0.01, 0.05, 0.1). Utilizing the MTS assay, viabilities were measured after 24h, 48h and 72h post-infection. Results depict two independent experiments, both conducted with triplicates per condition. Means +/- standard deviations are shown.

Both, mouse PDAC and primary pancreatic cells (PPC) cells were infected with rVSV-GFP at increasing MOIs, ranging from 0.001 to 0.1. In order to measure the viability, MTS assays were conducted from cells at 24h, 48h and 72h post infection. As illustrated in Figure 5, a decrease in cell viability over time was observed for the three different PDAC cell lines 110299, 530202 and 14128. The anti-tumoral effect positively correlated with the amount of virus the cells were infected with, as well as the time-point post-infection. All three cell lines demonstrate a similar trend in loss of viability. Importantly, no significant effects on cell viability could be observed in PPC cells in response to VSV infection, regardless of the MOI or time-point.

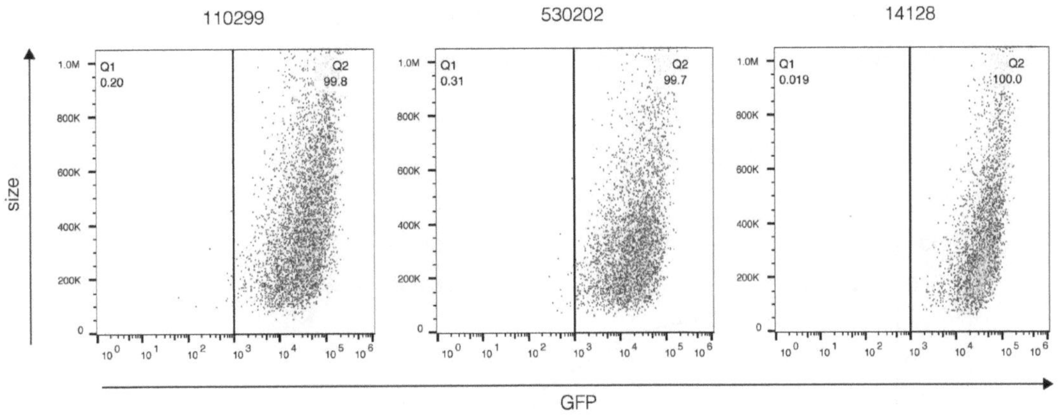

Figure 6: VSV infects PDAC cells with an efficacy of nearly 100%
Three different mouse PDAC cells lines, 110299, 530202 and 14128, were infected with rVSV-GFP at an MOI of 0.01. Utilizing flow cytometry, GFP-expressing cells were quantified at 24h, 48h and 72h post-infection. Representative dot plots are illustrated, depicting the number of GFP positive cells 24h post infection for each cell line. No significant changes were observed at the 48 and 72 hour time-points.

To further ensure that these effects are due to productive viral infection, we sought to determine the infection efficacy of rVSV-GFP on PDAC cells using GFP as a reporter. PDAC cell lines were infected with rVSV-GFP at an MOI of 0.01. Utilizing FACS, the percentage of GFP-expressing cells were quantified at 24h, 48h and 72h post-infection. Interestingly, the percentage of GFP-positive cells was consistent over the time monitored in all cell lines., Representative data are presented in Figure 6, which shows GFP expression at the 24h time-point for the three cell lines as an indicative result. Each dot plot is divided into two different gates, Q2 represents the amount of GFP positive cells, accounting for at least 99.7% of the cells.

4.2 S3I-201 causes reduction of PDAC cell viability *in vitro*

In addition to VSV, it was important to check if the STAT3 inhibitor S3I-201 has an effect in the PDAC cell lines as well. To this end, the cell lines, 110299, 530202 and 14128, as well as PPCs, were treated with increasing concentrations of the inhibitor over time for up to 72 hours. S3I-201 was titrated to determine the effects of different concentrations, in order to determine the optimal working concentrations for further experiments. Viabilities were compared by MTS assays.

As shown in Figure 7 below, the STAT3 inhibitor S3I-201 has antitumoral effects on the three tumor cell lines. While the STAT3 inhibitor concentration is directly correlated with the efficacy of killing the tumor cells, the three different cell lines show different response to the time points. For example, 72 hours are required for S3I-201 for an 80% reduction of viability in the 110299 cells, while this effect can already be achieved by 48 hours in 14128 cells, and as early as 24 hours for the 530202 cells. However, STAT3 inhibition caused no observable changes in viability in PPC cells, irrespective of the dose or time-point.

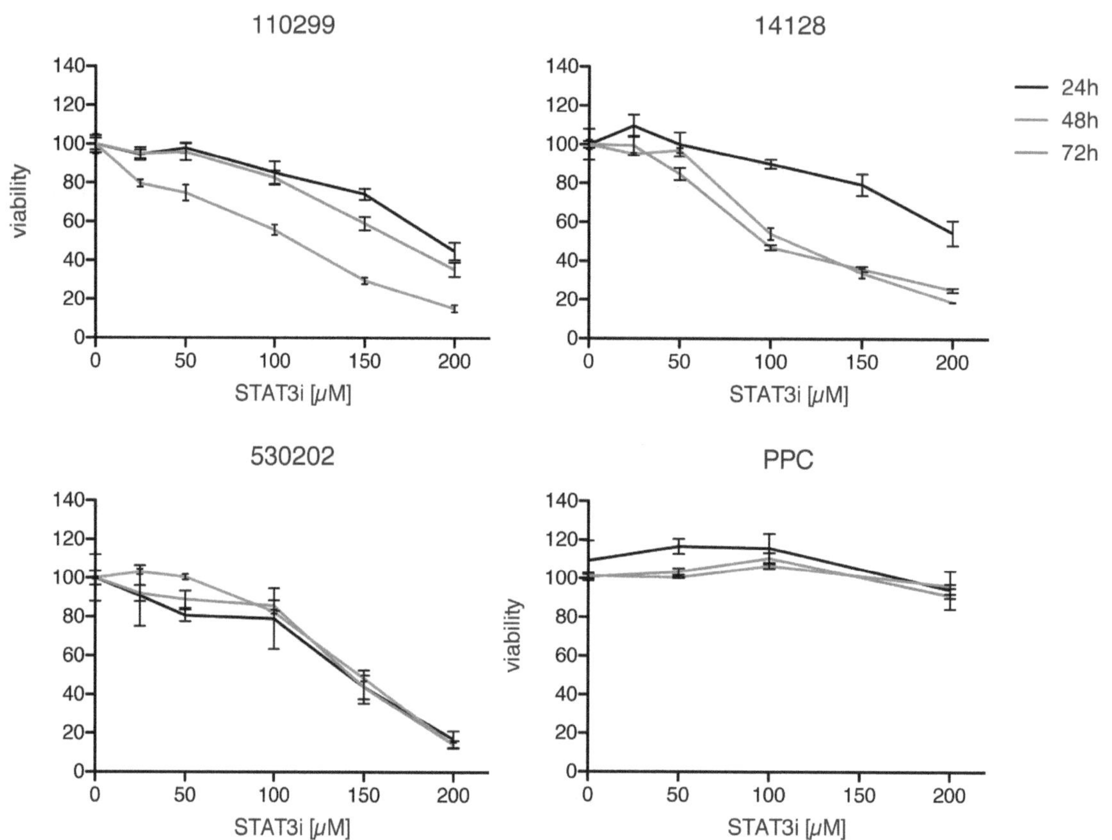

Figure 7: STAT3i shows anti-tumoral effect while leaving PPCs intact
Three different PDAC cells lines, as well as PPCs, were treated with the STAT3 inhibitor, S3I-201, in increasing concentrations (25µM, 50µM, 100µM, 150µM and 200µM). Utilizing the MTS assay, viabilities were measured after 24h, 48h and 72h post-treatment. Results depict two independent experiments, both conducted with triplicates per condition. Means +/- standard deviations are shown.

4.3 Combination therapy enhances PDAC killing and safety

The main aim of this project was to examine the combinational effect of VSV together with the STAT3 inhibitor, S3I-201. To investigate this, the inhibitor was titrated against different MOIs of rVSV-GFP, and cell viability was determined by MTS assays at different time-points post-treatment.

As depicted in Figure 8, we were able to detect a combinatorial effect, as evidenced by an enhanced anti-tumoral effect on PDAC cells when treated with both agents. The reduction of viable cells is again positively correlated with time, inhibitor concentrations and viral MOIs. Also the trend is similar among the three different cell lines. After 24h, PDAC cells still display around 30-80% healthy cells, depending on the treatment conditions. However, by 48h, the tumor cells reach a reduction in viability of below 10%. The lethal dose at which 50% of the cells are dead (LC 50) was defined as an MOI of 0.01 VSV and 100 µM S3I-201 for all cell lines, and these doses were used in all subsequent experiments.

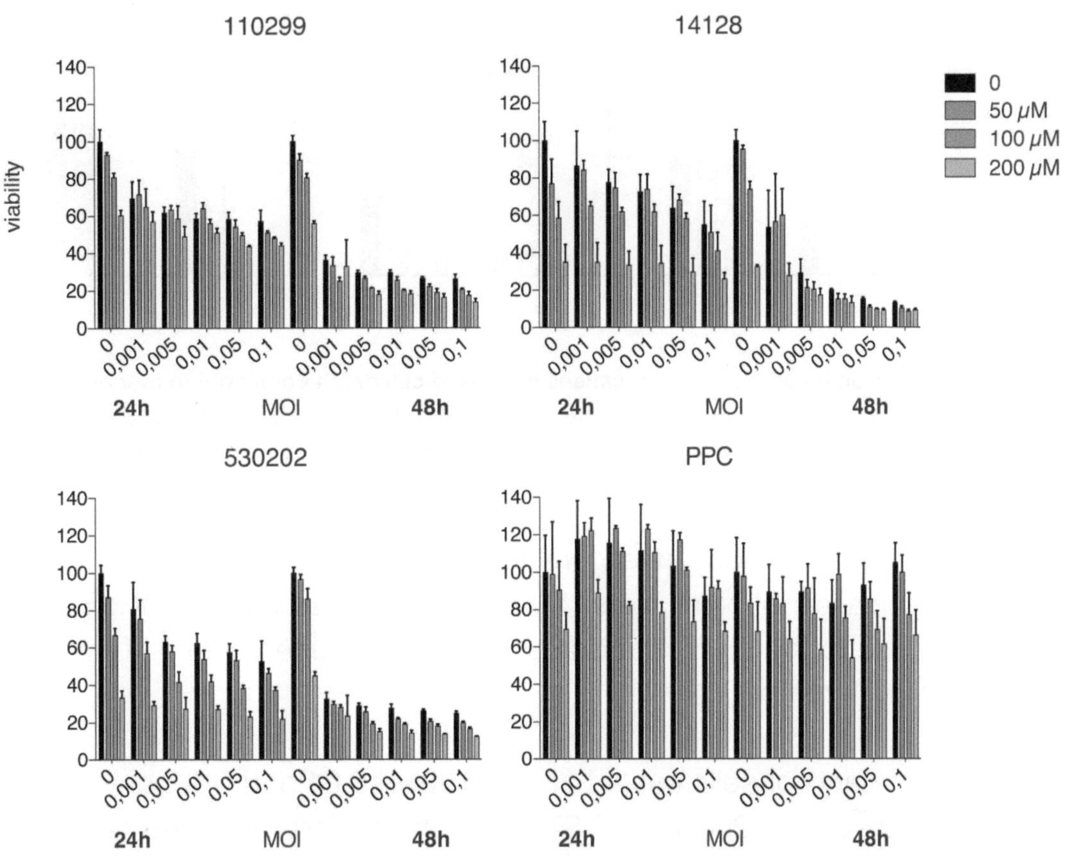

Figure 8: The combination therapy of rVSV with STAT3 inhibition leads to dose responsive anti-tumor effects
Three different PDAC cells lines, as well as PPCs, were treated with the STAT3 inhibitor S3I-201 in increasing concentrations (50µM, 100µM and 200µM) and combined with infections with rVSV-GFP at increasing MOIs (0.001, 0.005, 0.01, 0.05, 0.1). Viabilities were measured by MTS assay 24h and 48h post-treatment. Results depict two independent experiments, both conducted with triplicates per condition. Means +/-standard deviations are shown.

In contrast, the combination therapy in general had negligible effects on primary pancreatic cells, although the combinational therapy with 200 µM of the inhibitor seems to have a slight toxic effect on the healthy cells. However, at the conditions of the , PPCs seem barely affected and even after two days of treatment, they display viabilities above 80%.

To confirm these findings, cells were treated under the same conditions and stained with 7-Aminoactinomycin- D (7-AAD) for flow cytometric analysis. The results are depicted in Figure 9. For all three cell lines, no detectable increase in dead cells was observed in response to the STAT3 inhibitor. In contrast, for at least two of the three cell lines, we observed a significant increase in cell death upon treatment with rVSV at an MOI of 0.01, which positively correlated with time. However, the combinational therapy resulted in the highest percentage of cell death. By 48h, we observed a significant increase in dead cells for all three cell lines. Interestingly, the efficiency of the cell killing, and the rate at which this occurred, varied for the 3 cell lines.

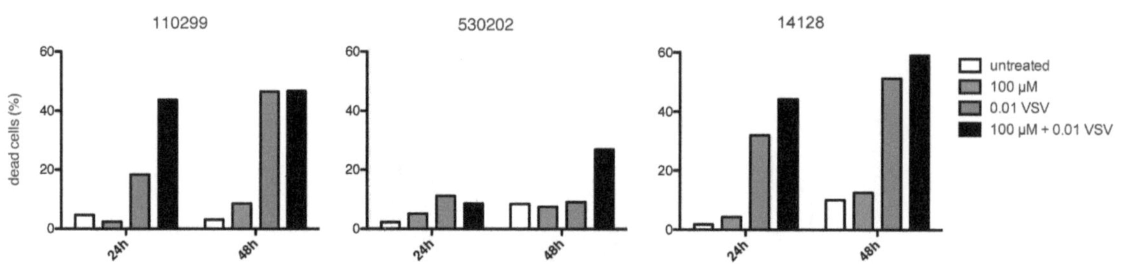

Figure 9: Combinational therapy causes increased cell death compared to monotherapy
Three different PDAC cells lines were treated with 100 µM of the STAT3 inhibitor, S3I-201, with VSV at a MOI of 0.01, or with both together. Utilizing a 7-AAD staining, percentage of dead cells were measured after 24h and 48h post-treatment by flow cytometry.

An important follow-up experiment, when combining virus with drugs, is the determination of viral titers, to rule out an attenuation of viral replication in response to the drug. To address this issue, supernatants of the same samples were analyzed by $TCID_{50}$.

Supernatants of cells treated at the MOIs of 0.01 and 0.1, combined with either 0 µM, 100 µM or 200 µM of STAT3 inhibitor were analyzed. Although the titers produced after infection at MOI 0.01 versus 0.1, as well for the different time-points, were not substantially different, viral titers in the PDAC cells were directly associated with time and infected MOI (Figure 10). Specifically, the viral titers for the tumor cells range between 10^8 and 10^{11}. However, these results suggest that S3I-201 has no effect on viral titers, as they stay the same irrespective of combined STAT3 inhibitor concentration. On the other hand, PPCs show significantly lower titers in the presence of the STAT3 inhibitor. Although they are generally lower, even without the addition of S3I-201, it seems that the inhibitor concentration causes a further decrease of viral titers in a dose-responsive manner. Hence, with 100 µM S3I-201 the titers range around only 10^1 - 10^3, while it is remarkably that in combination with 200 µM STAT3 inhibitor, the titers are below 10 02 on day one, and even not detectable anymore on day two.

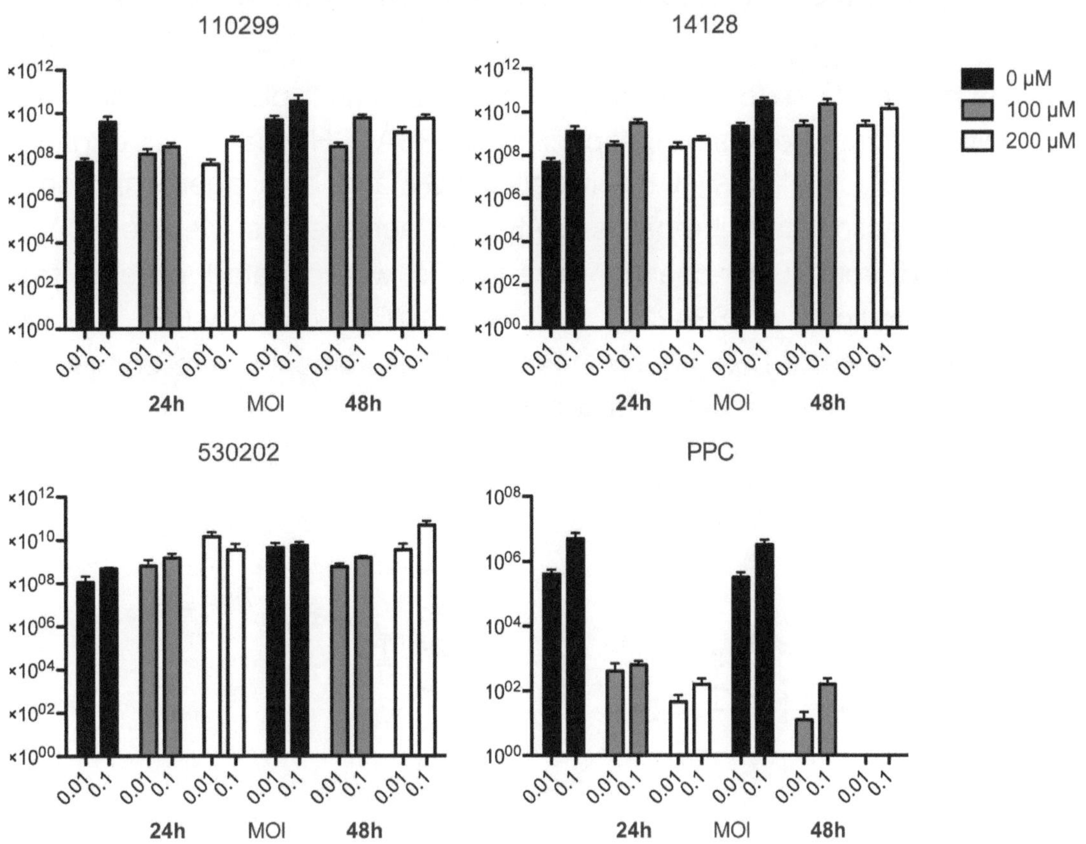

Figure 10: S3I-201 reduces the titers in PPCs while not affecting it in PDACs

Three different PDAC cells lines, as well as PPCs, were treated with the STAT3 inhibitor, S3I-201, in increasing concentrations (50µM, 100µM and 200µM), combined with infections with rVSV-GFP at increasing MOIs (0.01 and 0.1). Titers were measured by TCID$_{50}$ assay from supernatants collected at 24h and 48h post-treatment. Results depict two independent experiments, both conducted with quadruplicates per condition. Means +/- standard deviations are shown.

In order to confirm that these effects are attributable to the inhibitory effect of S3I-201, the same samples were taken to produce lysates for the detection of pSTAT3 Y705 utilizing Western blotting. As depicted in Figure 11, treating PDAC cells with 100 µM STAT3 inhibitor efficiently abrogated the presence of pSTAT3 Y705, regardless if in combination with rVSV-GFP or as single treatment. In contrast, pSTAT3 Y705 was detected in untreated samples and cells treated with only rVSV-GFP.

Figure 11: S3I-201 effectively abrogates pSTAT3 Y705 in PDAC

The PDAC cell line 14128 was treated with 100 µM S3I-201, infected with rVSV-GFP at an MOI of 0.01, or treated with a combination of both (100 µM + MOI 0.01). A specific antibody (1:200) was used for detection of the pSTAT3 Y705.

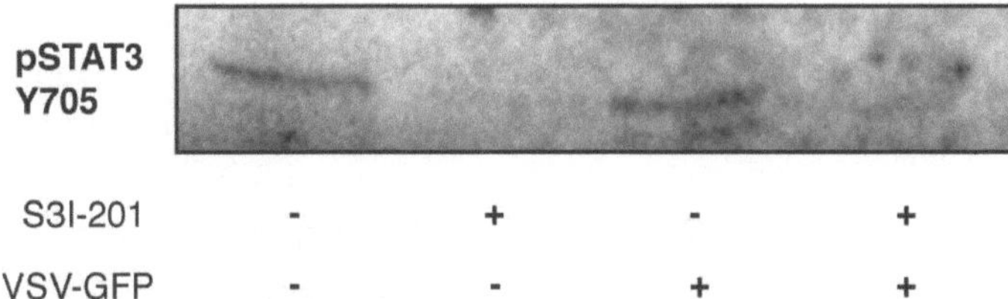

4.4 Combinational therapy in the microenvironment

As previously stated, the tumor microenvironment plays a crucial role in cancer initiation and progression, especially for PDAC. Hence, it was of great interest to elucidate the effects of the inhibitor, VSV and both agents in combination on cancer associated fibroblasts (CAFs). To achieve this, we used CAFs that were isolated from tumor-bearing mice. CAFs were treated either with increasing concentrations of S3I-201, increasing MOIs of rVSV-GFP, or with various combinations of both agents. Viability measurements were conducted after 24h and 48h post-treatment by MTS assay.

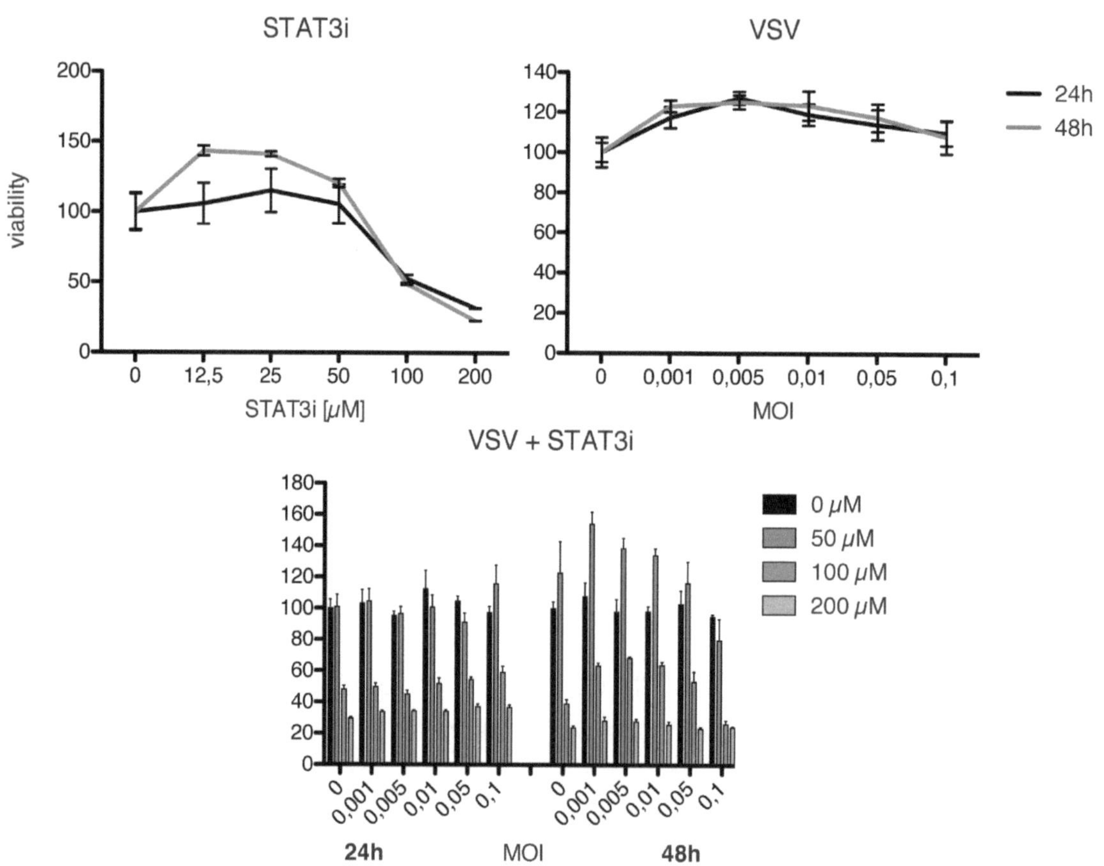

Figure 12: S3I-201 leads to a drastic decrease in CAF viability, whether as single treatment or in combination

Cancer associated fibroblasts (CAFs), isolated from tumor-bearing mice, were treated with the STAT3 inhibitor, S3I-201, in increasing concentrations (12.5µM, 25µM, 50µM, 100µM and 200µM) and / or infected with rVSV-GFP in increasing MOIs (0.001, 0.005, 0.01, 0.05, 0.1). Viabilities were measured by MTS assay at 24h and 48h post-treatment. Results depict two independent experiments, both conducted with triplicates per condition. Means +/- standard deviations are shown.

Figure 12 shows the effect of both single and combinational treatments of S3I-201 and rVSV-GFP on CAFs. While the cells are relatively insensitive to the STAT3 inhibitor at low concentrations, after 24h we can observe a drastic decrease in the viability in response to concentrations of 100 µM and above. On the other hand, CAF viability is not altered within the presence of rVSV infection, regardless of the utilized MOI. For example, even with an MOI of 0.1, fibroblasts display the same viabilities as untreated controls. More interestingly, while fibroblasts treated with only virus or combined with 50 µM S3I-201 did not experience a decrease in viability, samples from combinations containing either 100 µM or 200 µM displayed a crucial decline in viability to as low as 20%. Of note, at 48h post-treatment, when fibroblasts treated with 50 µM and 100 µM showed a slight increase in viability compared to the 24h time point.

In order to further investigate the relative ineffectiveness of VSV in CAFs, we conducted another experiment using the same conditions, in which cells were analyzed by flow cytometry. Contradictory to our expectations, we found VSV-mediated GFP expression in CAFs with an efficacy of 75% after a 24 hour infection, as illustrated in Figure 13. In addition, over 67% of cells expressed GFP after the combination therapy.

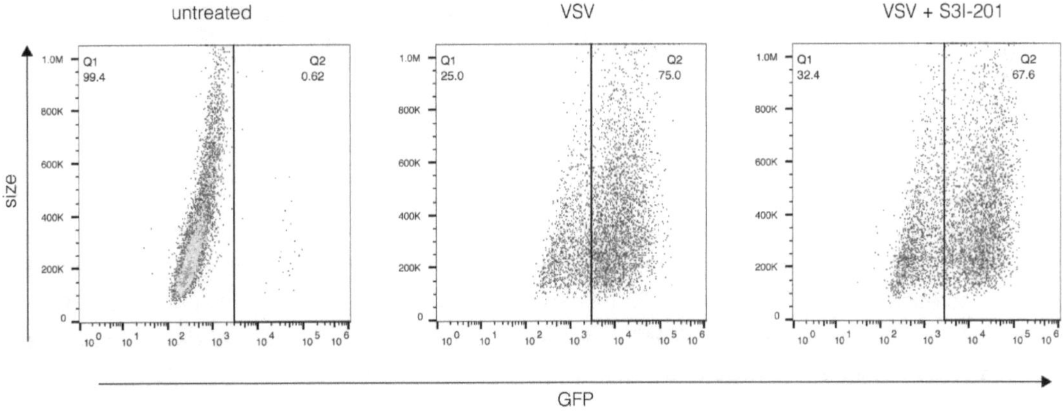

Figure 13: VSV infects CAFs cells with an efficacy of 75%
CAFs were infected with VSV at an MOI of 0.01, either alone, or in combination with 100 µM S3I-201. Utilizing flow cytometry, VSV-mediated GFP levels were measured after 24h post-infection. Representative dot plots are shown.

Again, it was important to ensure if these effects were due to the effectiveness of the STAT3 inhibitor. As previously described in the PDAC cells, S3I-201 effectively abrogates pSTAT Y705 also in the CAFs, shown in Figure 14. In the presence of the STAT3 inhibitor, the protein was not detectable anymore. Consistently, CAFs showed pSTAT3 Y705 bands when lysates were either left untreated or infected with rVSV-GFP.

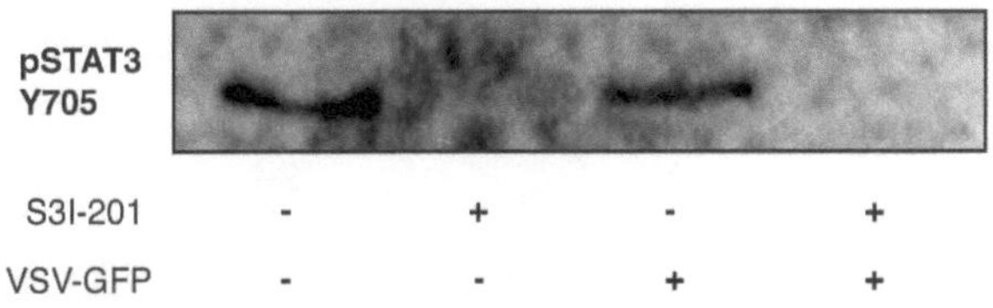

Figure 14: S3I-201 effectively abrogates pSTAT3 Y705 in CAFs
CAFs were treated with 100 µM S3I-201, infected with rVSV-GFP at an MOI of 0.01, or treated with a combination of both (100 µM + MOI 0.01). A specific antibody (1:200) was used for detection of the pSTAT3 Y705.

4.5 Effects of S3I-201 and VSV in co-culture

In addition to investigating the effects of S3I-201 and VSV on PDACs and CAFs alone, it was interesting to see whether a direct interaction of both cell types would have an impact on the outcome of the treatments. Utilizing trans-well plates, we were able to culture PDACs together with CAFs on a membrane, ensuring that the cells were not in direct contact with each other, but could still exchange molecules to enable cross-talk among them. After incubation for 24 hours post-treatment, viabilities were measured by MTS assay on both cell types individually.

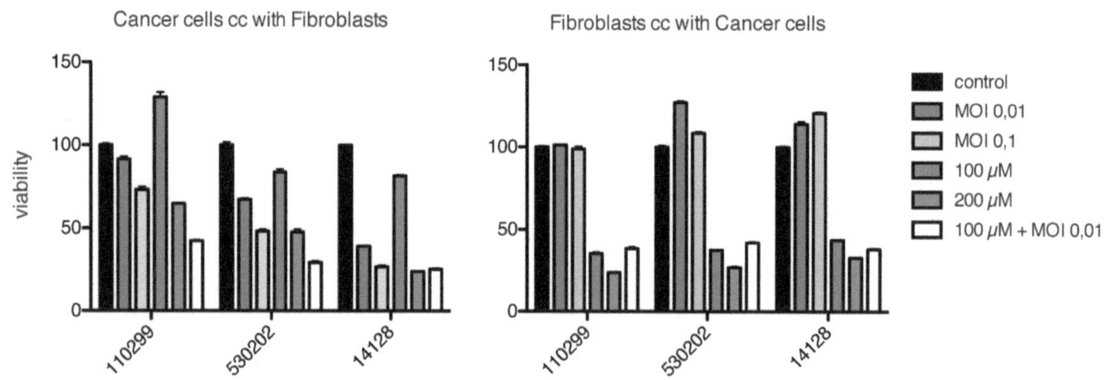

Figure 15: Combinational therapy is effective in co-coltural models
Three different PDAC cells lines were co-cultured with CAFs in trans-well plates. Co-cultures were treated with the STAT3 inhibitor, S3I-201, in increasing concentrations (100µM and 200µM,), infected with rVSV-GFP at increasing MOIs (0.01 and 0.1), or treated with a combination of both (100 µM + MOI 0.01). Viabilities were measured in each cell type by MTS assay at 24h post-treatment. Results depict triplicates per condition. Means +/- standard deviations are shown.

All three PDAC cell lines were tested in combination with CAFs, as shown in Figure 15 above. On the left, cell viabilities of cancer cells, which were co-cultured with CAFs, are shown. As previously described, we observed a decrease in viabilities in response to both, STAT3 inhibitor and VSV, in a dose-responsive manner. Again, the strongest cytotoxic effect was observed for the combination therapy, resulting in 60-80% cytotoxicity in all PDAC cell lines. However, these results are not fully consistent with fibroblasts that were co-cultered with PDACs. VSV infections again had no measurable effect on CAF viability, while S3I-201 treatment was again quite effective, resulting in a reduction in viability to up to 15%. The combination therapy did not provide any additional benefit compared to STAT3 inhibitor alone.

4.6 Combination therapy in 3D

In order to take the next step from bench to bedside, we were aiming to confirm in cell culture systems which better reflect the complexity in reality. Hence, we investigated our treatment approach in 3D.

As a first step, we tried to measure viability on tumor spheroids. Again, increasing STAT3 inhibitor concentrations and increasing MOIs of rVSV-GFP were added to 530202 cells, cultured as a spheroid, for up to 72h hours. Viabilities were determined by MTS assay. As shown in Figure 16, neither S3I-201 nor rVSV-GFP affected the measured viability of the tumor spheroids, regardless of concentration, MOI and time. All time-point showed viabilities around 100%.

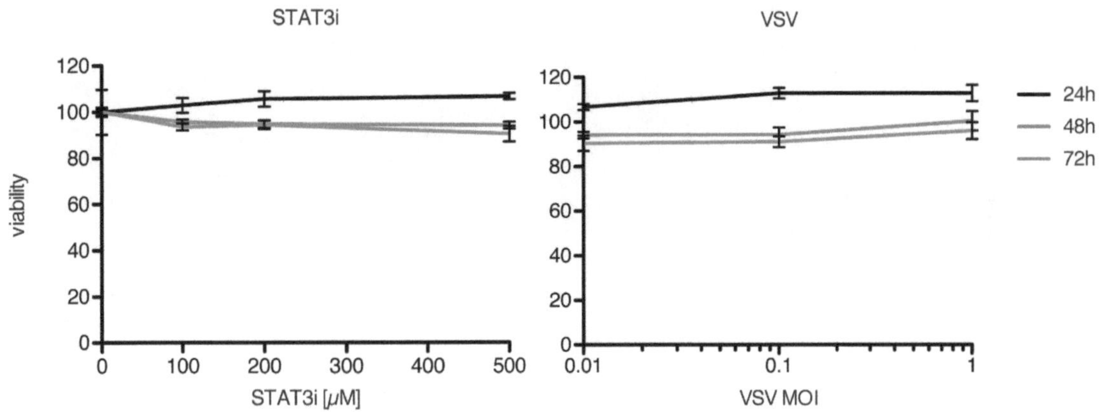

Figure 16: PDAC tumor spheroids neither respond to S3I-201 nor to rVSV-GFP
5 x 10^3 cells of the PDAC cell line 530202 were seeded into an ultra-low attachment 96 well plate in order to generate homogenous tumor spheroids, which were treated with the STAT3 inhibitor, S3I-201, in increasing concentrations (100µM, 200µM and 500µM) or infected with rVSV-GFP in increasing MOIs (0.01, 0.1 and 1). Utilizing the MTS assay, viabilities were measured after 24h, 48h and 72h post-treatment. Results triplicates per condition. Means +/- standard deviations are shown.

However, taking a look at the tumor spheroids under the microscope, we saw crucial differences in their morphology (Figure 17). Untreated spheroids maintained their morphology over 72h and stayed unaffected. In contrast, spheroids treated with S3I-201, rVSV-GFP or both displayed decisive differences. While untreated spheroids were well rounded with defined borders, treated spheroids showed a loss in their shape and disrupted borders. These effects positively correlated with time, as well as the combinational treatment caused these effects stronger than single treatments, resulting in fully disrupted spheroids.

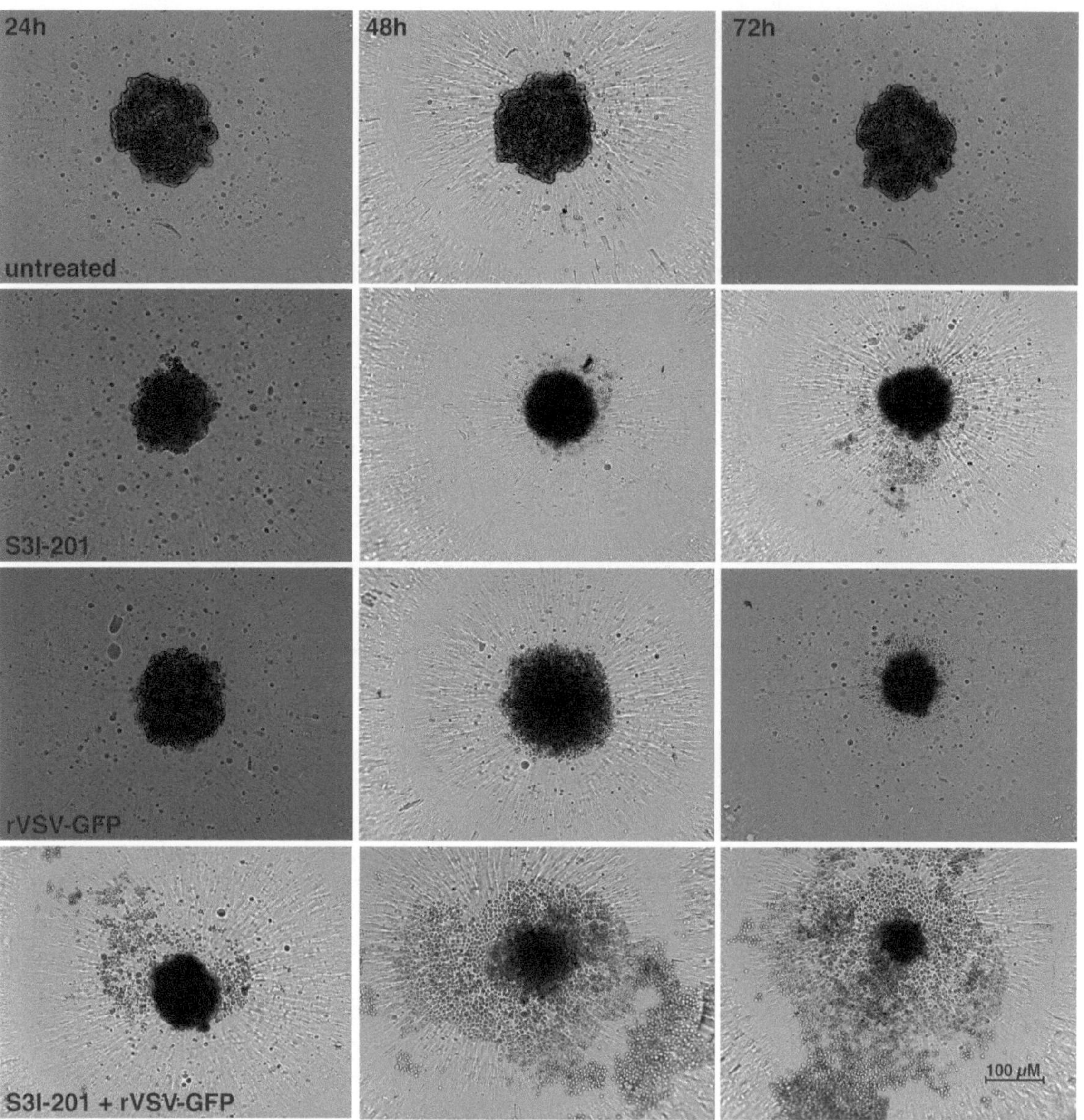

Figure 17: Combinational treatment disrupts the spheroidal morphology

5 x 10³ cells of the PDAC cell line 530202 were seeded into an ultra-low attachment 96 well plate in order to generate homogenous tumor spheroids, which were treated with 500 µM STAT3 inhibitor, infected with rVSV-GFP at an MOIs of 1 or both. Utilizing a live cell movie analyzer, images were acquired after 24h, 48h and 72h post-treatment.

As an additional experiment, we tried to investigate these effect in 3D on are more molecular level. In order to achieve this, we cultured 530202 cells for an appropriate amount of time in matrigel, until they formed homogenous spheroids. Then, spheroids were left untreated or were treated with S3I-201, rVSV-GFP or both for 48 hours. After fixation and staining of the spheroids, they were analyzed on a confocal microscope, depicted in Figure 18.

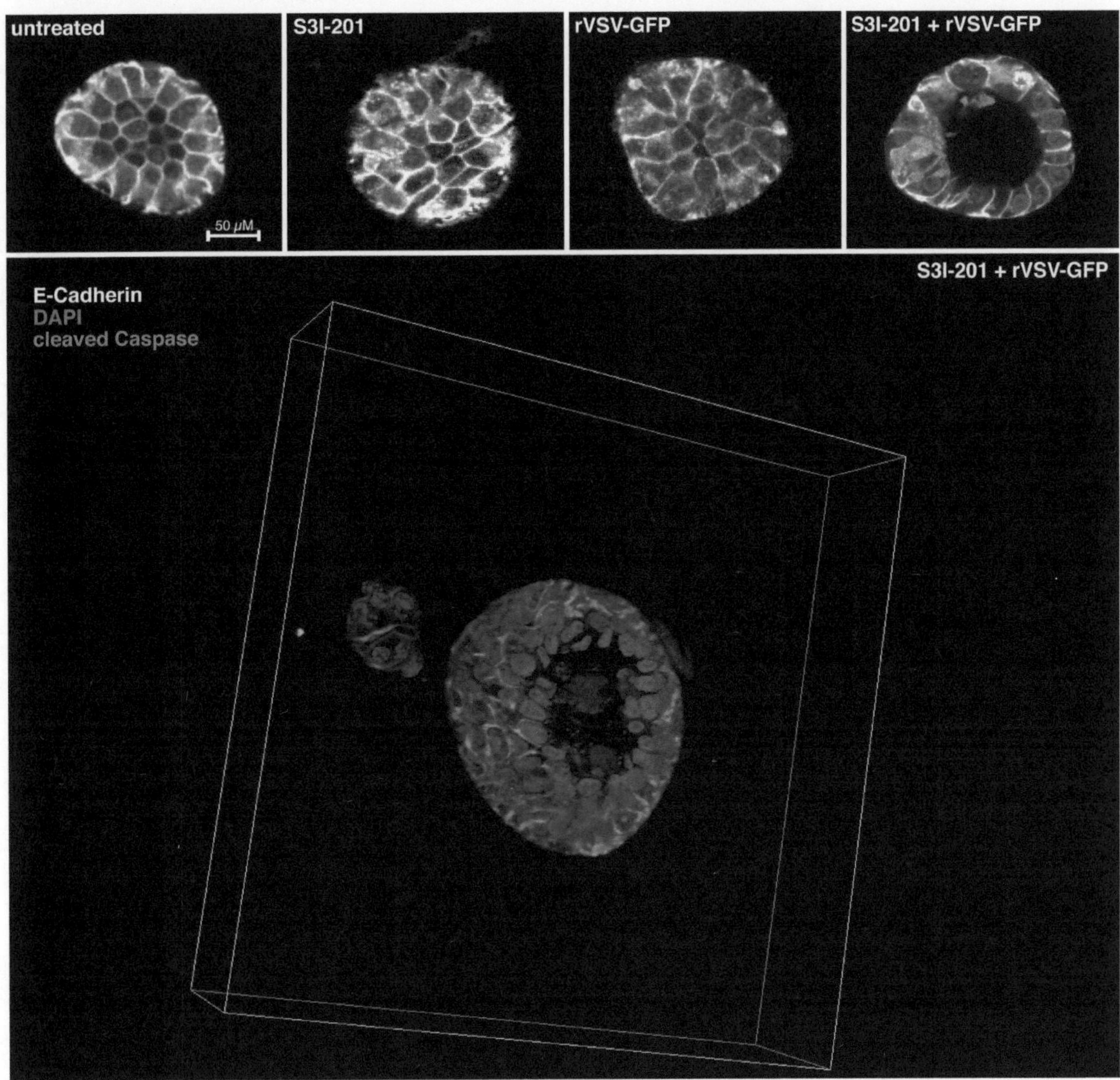

Figure 18: PDAC tumor spheroids undergo apoptosis upon STAT3 inhibition and VSV infection
530202 PDAC cell line was seeded in matrigel for 2-3 days in order to generate homogenous tumor spheroids, which were treated with 100 µM S3I-201, infected with rVSV-GFP at an MOI of 0.01 or both. After fixation, spheroids were DAPI stained (blue) as well as against E-Cadherin (grey) and cleaved Caspase (red). Spheroid pictured were acquired using confocal microscopy.

As indicated in red, cells that were treated underwent apoptosis. However, interesting differences were observed upon the different treatments. Spheroids treated with S3I-201 displayed apoptosis rather in the inner of the cell mass, whereas virus treated spheroids displayed apoptotic zones at the border. Contrary to that, combinational treatment was not just observed throughout the whole spheroid, also the signal of cleaved Caspase seemed to be increased. Also the 3D reconstruction of a image stack confirmed, that apoptotic zones were detectable in the inner and on the border of the spheroid.

4.7 Inhibition of migration

Based on various previous reports, we hypothesized that an inhibition of STAT3 would lead to a significant decrease in the ability of PDAC and stromal cells to migrate. We, therefore, sought to visualize the migratory ability of both cancer cells and fibroblasts under various conditions. In order to achieve this, we cultured the cells until they reached approximately 100% confluency. This was followed by scratching the confluent layer, in order to monitor if the "wound" would be healed, and if so, how fast. Cells were monitored over the course of 72h, with images being captured at five minute intervals. The resulting stack of images was assembled into a movie (Appendix pdf-version). Screen shots of the 0 time-point, as well as the time-point at which the wound was closed, are, illustrated in Figure 19 and 20 below.

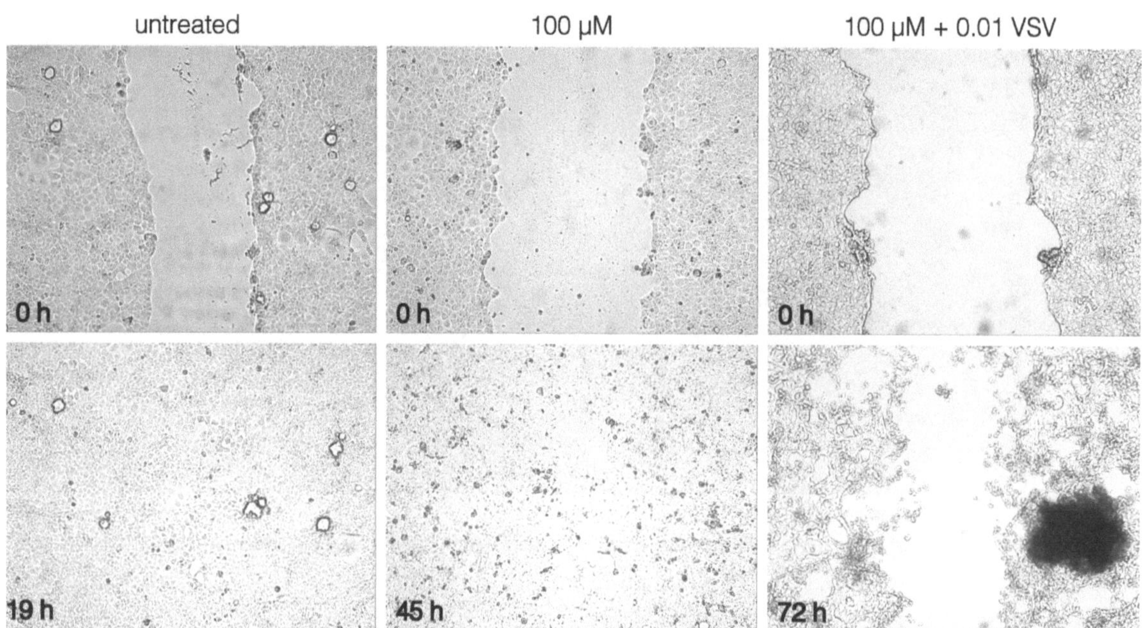

Figure 19: Inhibition of STAT3 impairs the ability to migrate in PDAC cells
The PDAC cell line, 530202, was cultured until 100% confluency, and the monolayer was scratched using a 200 µl pipette tip. Afterwards, cells were either left untreated, treated with 100 µM S3I-201, or treated with both the inhibitor and rVSV-GFP at an MOI of 0.01. Images were captured every 5 minutes utilizing a cell movie analyzer for 72h. Screen shots were taken at the indicated time-points.

As shown in Figure 19, treatment with S3I-201, either alone or in combination with rVSV-GFP, significantly influenced the migratory/proliferative potential of the PDAC cell line, 530202. Cells that were left untreated were able to grow back together to a confluent layer within only 19 hours. In contrast, cancer cells that were treated with S3I-201 were drastically delayed in their wound-healing capacity requiring 45 hours to reach a 100% confluent state. Even more striking, were the effects of the combination treatment. Cells that were treated with the inhibitor and VSV together, were not only completely impaired in their ability to close the wound, but after 72 hours, the cells demonstrated a cytopathic effect and seemed to no longer viable based on their morphology.

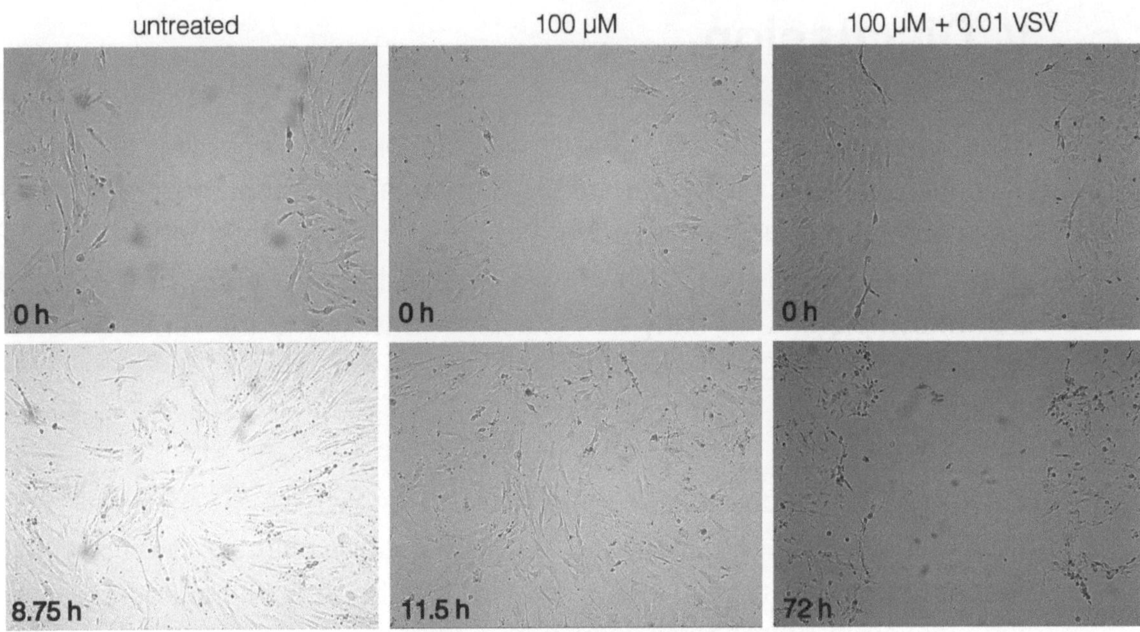

Figure 20: Inhibition of STAT3 impairs the ability to migrate in CAFs
CAFs were cultured until 100% confluency, and monolayers were scratched by a 200 µl pipette tip. Afterwards, cells were either left untreated, treated with 100 µM S3I-201, or treated with both the inhibitor and VSV at an MOI of 0.01. Images were captured every 5 minutes utilizing a cell movie analyzer for 72h. Screen shots taken at the indicated time-points are shown.

These finding also apply to the migratory and proliferative potential of CAFs, depicted in Figure 20. Untreated CAFs showed an even improved „wound" healing potential compared to PDAC cells and grew back together already within 8.75 hours. Although not as effective as in PDACs, treatment with the STAT3 inhibitor again delayed this capability to 11.5 hours. In contrast, the combination therapy again completely abrogated the wound healing effect and led to killing of the CAFs, based on their morphology.

V. Discussion

Although huge efforts have been made in order to improve the prognoses of patients diagnosed with pancreatic ductal adenocarcinoma, this disease still ranks among the most aggressive and deadliest types of cancer (Rossi, Rehman et al. 2014). Since prevention of this disease has proven to be difficult due to a lack of symptoms and early metastasis, in addition to the rapid development of resistances, PDAC is unequivocally in urgent need of novel therapeutic approaches (Kleeff, Korc et al. 2016).

It is known that oncolytic VSV (Hastie and Grdzelishvili 2012) as well as inhibition of STAT3 (Sharma NK 2014) are both potentially effective treatment options in cancer therapy. It was even shown for hepatocellular carcinoma that a combination of both agents results in synergistic effects, not only in vitro, but in vivo as well (Marozin, Altomonte et al. 2015). Aiming to establish this therapeutic approach in pancreatic cancer in our lab, this study suggests that a combinational treatment of these two agents is also effective in PDAC.

Our findings suggest that VSV and the STAT3 inhibitor S3I-201 have anti-tumor effects in PDAC cells, and that a combinatorial treatment results in improved efficacy compared to either monotherapy.
Utilizing flow cyometry, we were able to show that rVSV-GFP infects three different PDAC cell lines with an efficacy of approximately 100% (Fig. 6). Aiming to measure the tumor cell killing effects of both therapeutic agents, we conducted MTS assays. This technique is based on the conversion of MTS to formazan, which is implemented by NADH, a molecule that is produced by dehydrogenases solely in metabolically active cells. Hence, this approach measures the formazan concentration in the medium and directly links metabolic activity to viability of the cells.
Based on this approach, both, rVSV-GFP and S3I-201, significantly reduced the viability of PDAC cells (Figure 5 & 7). Concentrations of the inhibitor, MOIs of the virus, as well as the time-period of the treatment positively correlated with their anti-tumoral effect. The fact that some of the cell lines responded earlier than others is attributable to the naturally occurring intratumoral heterogeneity. While PDAC cells were harmed, healthy PPC cells stayed fully viable after both of the treatments. These results were consistent with our expectations, since previous findings suggested the up-regulation of pSTAT3 (Corcoran, Contino et al. 2011) as a therapeutic target. In addition, we were able to show the high susceptibility of PDAC cells to rVSV-GFP, which was expected based on previous studies (Filley and Dey 2017).

More interesting was the effect of both agents as an improved therapeutic approach. Again based on the findings of our viability assays (Figure 8), we observed that both, virus and inhibitor, decreased the viability of the tumor cells. But, with a combination of both, we even saw an improved cytotoxic effect and were able to reduce the viability of our PDAC cells to below 10%. However, in testing these combinations on our PPCs, we detected a slight cytotoxic effect in combinations with a high dose (200 µM) of STAT3 inhibitor, which reduced the viability to 60%. To confirm these finding, a dead cell staining via 7-AAD was conducted and quantified using flow cytometry (Figure 9). Although the three PDAC cell lines again responded differently in terms of the amount of time required to reach a similar effect, the significant increase in the dead population with both agents compared to single therapies confirmed the hypothesis that a combinational treatment improves the cytotoxic effect in PDAC cells. To exclude that the effect was the consequence of nonspecific effects of the STAT3 inhibitor, Western blot analysis was performed, which confirmed that the inhibitor, with or without rVSV-GFP, successfully abrogated the protein levels of pSTAT3 Y705, while untreated and infected cells showed constitutive activation (Figure 11), suggesting that the therapeutic effects S3I-201 may indeed be attributable to inhibition of STAT3 activation. In order to fully rule out any off target effects of the inhibitor, one would need to utilize a more specific approach, such as siRNA against STAT3.

In addition, the investigation of the viral titers after the treatments confirmed that the STAT3 inhibitor does not interfere with viral replication in the tumor cells (Figure 10). It was crucial to ensure that the combination therapy would not inadvertently lead to rVSV-GFP attenuation in tumor cells. More interesting is the effect of the combination therapy on healthy PPCs. While they already produce naturally lower titers due to their intact interferon response, we observed that the STAT3 inhibitor further reduced the viral titers in the primary cells. Furthermore, the ability of S3I-201 to inhibit rVSV-GFP replication in healthy cells is dose-responsive. Therefore, the combination therapy not only leads to enhancedd cytotoxic effects in PDAC cells, but it improves the safety of oncolytic virus therapy by suppressing viral replication exclusively in healthy tissues.
We speculate that these finding are ascribable to the antagonistic effect of STAT1 and STAT3. Recent studies found that STAT1 acts as an anti-viral protein by the induction of the IFN-type I pathway. Simultaneously, STAT3 acts as a pro-viral protein by negatively regulating STAT1 (Ho and Ivashkiv 2006). Hence by inhibiting STAT3 in combination with the oncolytic virus, rVSV-GFP, we believe that the safety is improved by enhancing the anti-viral effect in healthy tissues. At the same time, since the constitutive activation of STAT3 in PDAC cells is well known for its tumor supportive effect, S3I-201 and rVSV-GFP together improve their anti-tumoral effect in an additive manner.

Even more promising are our findings in three-dimensional cell culture systems, demonstrating that the efficacy of the combination therapy is not limited to 2D. Using ultra-low attachment plates, we were capable of generating tumor spheroids that were highly homogenous in size and shape (Figure 17). Based on their morphology, untreated spheroids maintained their size and shape, indicating that they maintained their viability. In contrast, spheroids treated with either S3I-201 or rVSV-GFP began to lose their shape and borders lost their definition, and they shrank over time. The combination therapy even enhanced these effects, leading to complete degradation of the spheroids. We suggest that these effects are a reflection of the loss of cell viability, leading to the inability of proper cell-cell-interactions. However, we were not able to confirm the decrease in viability via a quantitative assay (Figure 16). According to MTS quantification, tumor spheroids remain fully viable, although this data was contradictory to the morphology of the spheroids. We believe that the viability of spheroids cannot be accurately quantified via MTS assay, since this approach is based on metabolic activity. Tumor spheroids are known to display metabolic activity solely in their outer zones, while towards to center, the cells become quiescent or necrotic (Zanoni, Piccinini et al. 2016). Furthermore, it is conceivable that the MTS solution does not sufficiently penetrate through the three-dimensional architecture, and the tumor spheroids have a reduced surface area than 2D cell cultures do.

Furthermore, we aimed to investigate the effects in 3D cell culture on a more molecular level. To achieve this, we investigated spheroids grown and treated in matrigel under the confocal microscope. Stained for cleaved caspase, we were interested in whether our treatments cause apoptosis in the 3D cultures (Figure 18). We found little apoptotic zones throughout the whole spheroid when treated with STAT3 inhibitor, while these zones were detectable solely on the borders when treated with rVSV-GFP. We hypothesized that this is attributable to the inability of the virus to penetrate the spheroids effectively. However, when treated with a combination of both agents, it does not just seem to enhance the apoptotic effect, also the 3D reconstruction of our image stack shows us that these zones were found both in the inner and outer portions of the spheroid, indicating that the combination therapy is cytotoxic in three-dimensional cell culture systems as well and predicting that it could also be effective in actual tumors.

In addition to our investigations on PDACs in 2D and 3D systems, we were interested in the response of the tumor microenvironment to the combinational therapy. As the PDAC microenvironment is particularly marked by the presence of stromal cells, we attempted to recreate the interplay between PDAC cells and cancer-associated fibroblasts in vitro. We began by determining the viability of CAFs after treatment with the inhibitor, rVSV-GFP or both (Figure 12). We found that although the fibroblasts were highly susceptible to S3I-201, they were highly resistant to the viral treatment. Nevertheless, combining the two agents significantly decreased the viabilities of the CAFs, although to a similar level as that achieved by the STAT3 inhibitor alone, indicating that the role of rVSV in CAFs is probably negligible. Furthermore, Western blot analysis revealed that these effects correlated with the concentration of inhibitor applied (Figure 14).

Co-culture experiments confirmed these findings (Figure 15). While CAFs that were grown in the presence of PDACs were killed by the inhibitor and the combination therapy, cells treated solely with virus did not respond, according to data obtained via MTS assay.

Although it was conjecturable, that this effect is due to the ineffectiveness of rVSV-GFP to infect CAFs, we obtained contradictory results by flow cytometric analysis (Figure 13). Even though the efficacies of infection were around 67-75% instead of 100% as in PDACs, we found that the virus was capable of infecting the majority of the cells.

Previously reported findings have described the cross-talk of cancer cells with CAFs through the IL-6 / STAT3 axis (Hendrayani, Al-Khalaf et al. 2014). Activated fibroblasts induce the pSTAT3 pathway in PDACs with secretion of IL-6, leading to various effects, like cancer progression and epithelial to mesenchymal transition. Reciprocally, pSTAT3 signaling results in a positive-feedback loop and further activates CAFs by up-regulation of smooth muscle actin and TGF-beta, as well as negatively regulating INK4A and p53.

We propose that the disruption of this cross-talk, via application of S3I-201, is the reason for the significant decrease in viability of the CAFs. Additionally, it has been shown that CAFs generally have an intact IFN-type I response (Hosein, Livingstone et al. 2015). Therefore, we assume that the immune response in our CAFs was just partially disrupted, allowing the virus to infect the cells while inhibiting their replication cycle. The reasons for our contradictory data in comparison to previous findings (Ilkow, Marguerie et al. 2015) are not entirely clear, and further investigations are warranted to elucidate the possible mechanisms. However, we assume that this could be attributable to the different PDAC models the CAFs were isolated from. Another potential explanation could be that rVSV-GFP infected CAFs underwent quiescence before initiating apoptosis, which would impede the quantification using MTS assay.

As an experiment in parallel, using a simple scratch assay, we investigated the effect of our therapeutic approach on the proliferative and migratory potential of both, the PDACs and CAFs (Figure 19 & 20). As we previously described, S3I-201 significantly decreased this potential and delayed the wound healing effect of the cancer cells by up to 26 hours, and approximately 3 hours for the fibroblasts. Even more striking were the effects with the combination therapy. Cells showed strong cytopathic effect and died rapidly, resulting in a unclosed scratch even after 72 hours.

We speculate that this effect is due to STAT3's crucial role in invasiveness and metastasis by promoting the migratory potential of cancer cells, as it was demonstrated in previous studies (Teng, Ross et al. 2014).

Taken together, we believe that combining the STAT3 inhibitor S3I-201 with rVSV-GFP displays a novel and promising therapeutic approach for treating pancreatic ductal adenocarcinoma. Our in vitro studies showed that combining these two anti-tumoral agents not only improve their effects in an additive manner, but also safety in healthy nontumor cells was enhanced. In addition, the killing effect on cancer-associated fibroblasts could not only decrease the crosstalk between CAFs and PDACs, also the spread of rVSV-GFP within the tumoral tissue could be improved. Besides the cytotoxic effects that counteract cancer progression and maintenance, the anti-proliferative and anti-migratory potential of this combinational approach even has the potential to inhibit the ability to metastasize. Since these findings are based on 2D, 3D and co-cultural approaches in vitro, we believe it is warranted to continue these investigations in vivo to further characterize and develop the approach for clinical application.

VI. List of Abbreviations

5-FU	5-Fluorouracil
7-AAD	7-Aminoactinomycin D
ADM	Acinar to ductal metaplasia
CAF	Cancer associated fibroblast
CM	Conditioned medium
CPE	Cytopathic effect
DMEM	Dulbecco's Modified Eagle Medium High Glucose
ECM	Extracellular matrix
EMT	Epithelial to mesenchymal transition
FACS	Fluorescence activated cell sorting
FDA	Food and drug administration
GM-CSF	Granulocyte macrophage colony stimulating-factor
IFN	Interferon
IL-6	Interleukin 6
ISRE	Interferon Sensitive Response Element
JAK	Janus activated kinase
LD_{50}	Lethal dose 50
MMP	Matrix metalloproteinase
MTS	3-(4,5-dimethylthiazol-2-yl)-5-(3-carboxymethoxyphenyl)-2-(4-sulfophenyl)-2H-tetrazolium
NM	Normal medium
OV	Oncolytic virus
PanINs	Pancreatic intraepithelial neoplasias
PBS	Phosphate buffered saline
PDAC	Pancreatic ductal adenocarcinoma
PFA	Paraformaldehyd
PKR	Protein kinase R
PPC	Primary pancreatic cells
PSC	Pancreatic stellate cell
rcf	Relative centrifugation force
RT	Roomtemperature
S3I-201	STAT3 inhibitor
Shh	Sonic hedgehog
STAT	Signal transducer and activator of transcription
$TCID_{50}$	Tissue culture infection dose 50
TGFβ	Transforming growth factor beta
ULA	Ultra low attachment
VSV	Vesicular stomatitis virus

VII. References

Ahmed, S., A. D. Bradshaw, S. Gera, M. Z. Dewan and R. Xu (2017). "The TGF-beta/Smad4 Signaling Pathway in Pancreatic Carcinogenesis and Its Clinical Significance." J Clin Med 6(1).

Altomonte, J. and O. Ebert (2012). "Replicating viral vectors for cancer therapy: strategies to synergize with host immune responses." Microb Biotechnol 5(2): 251-259.

Bardeesy, N. and R. A. DePinho (2002). "Pancreatic cancer biology and genetics." Nat Rev Cancer 2(12): 897-909.

Burris, H. A., 3rd, M. J. Moore, J. Andersen, M. R. Green, M. L. Rothenberg, M. R. Modiano, M. C. Cripps, R. K. Portenoy, A. M. Storniolo, P. Tarassoff, R. Nelson, F. A. Dorr, C. D. Stephens and D. D. Von Hoff (1997). "Improvements in survival and clinical benefit with gemcitabine as first-line therapy for patients with advanced pancreas cancer: a randomized trial." J Clin Oncol 15(6): 2403-2413.

Carr, R. M. and M. E. Fernandez-Zapico (2016). "Pancreatic cancer microenvironment, to target or not to target?" EMBO Mol Med 8(2): 80-82.

Chiocca, E. A. (2002). "Oncolytic viruses." Nat Rev Cancer 2(12): 938-950.

Conroy, T., F. Desseigne, M. Ychou, O. Bouche, R. Guimbaud, Y. Becouarn, A. Adenis, J. L. Raoul, S. Gourgou-Bourgade, C. de la Fouchardiere, J. Bennouna, J. B. Bachet, F. Khemissa-Akouz, D. Pere-Verge, C. Delbaldo, E. Assenat, B. Chauffert, P. Michel, C. Montoto-Grillot, M. Ducreux, U. Groupe Tumeurs Digestives of and P. Intergroup (2011). "FOLFIRINOX versus gemcitabine for metastatic pancreatic cancer." N Engl J Med 364(19): 1817-1825.

Corcoran, R. B., G. Contino, V. Deshpande, A. Tzatsos, C. Conrad, C. H. Benes, D. E. Levy, J. Settleman, J. A. Engelman and N. Bardeesy (2011). "STAT3 plays a critical role in KRAS-induced pancreatic tumorigenesis." Cancer Res 71(14): 5020-5029.

Darnell, J. E., Jr. (1997). "STATs and gene regulation." Science 277(5332): 1630-1635.

Feig, C., A. Gopinathan, A. Neesse, D. S. Chan, N. Cook and D. A. Tuveson (2012). "The pancreas cancer microenvironment." Clin Cancer Res 18(16): 4266-4276.

Fernandez, M., M. Porosnicu, D. Markovic and G. N. Barber (2002). "Genetically engineered vesicular stomatitis virus in gene therapy: application for treatment of malignant disease." J Virol 76(2): 895-904.

Ferro, R. and M. Falasca (2014). "Emerging role of the KRAS-PDK1 axis in pancreatic cancer." World J Gastroenterol 20(31): 10752-10757.

Filley, A. C. and M. Dey (2017). "Immune System, Friend or Foe of Oncolytic Virotherapy?" Front Oncol 7: 106.

Fukuhara, H., Y. Ino and T. Todo (2016). "Oncolytic virus therapy: A new era of cancer treatment at dawn." Cancer Sci 107(10): 1373-1379.

Goldufsky, J., S. Sivendran, S. Harcharik, M. Pan, S. Bernardo, R. H. Stern, P. Friedlander, C. E. Ruby, Y. Saenger and H. L. Kaufman (2013). "Oncolytic virus therapy for cancer." Oncolytic Virother 2: 31-46.

Hartkopf, A. D., T. Fehm, M. Wallwiener and U. Lauer (2012). "Oncolytic Viruses to Treat Ovarian Cancer Patients - a Review of Results From Clinical Trials." Geburtshilfe Frauenheilkd 72(2): 132-136.

Hastie, E. and V. Z. Grdzelishvili (2012). "Vesicular stomatitis virus as a flexible platform for oncolytic virotherapy against cancer." J Gen Virol 93(Pt 12): 2529-2545.

Heiber, J. F., X. X. Xu and G. N. Barber (2011). "Potential of vesicular stomatitis virus as an oncolytic therapy for recurrent and drug-resistant ovarian cancer." Chin J Cancer 30(12): 805-814.

Hendrayani, S. F., H. H. Al-Khalaf and A. Aboussekhra (2014). "The cytokine IL-6 reactivates breast stromal fibroblasts through transcription factor STAT3-dependent up-regulation of the RNA-binding protein AUF1." J Biol Chem 289(45): 30962-30976.

Ho, H. H. and L. B. Ivashkiv (2006). "Role of STAT3 in type I interferon responses. Negative regulation of STAT1-dependent inflammatory gene activation." J Biol Chem 281(20): 14111-14118.

Hosein, A. N., J. Livingstone, M. Buchanan, J. F. Reid, M. Hallett and M. Basik (2015). "A functional in vitro model of heterotypic interactions reveals a role for interferon-positive carcinoma associated fibroblasts in breast cancer." BMC Cancer 15: 130.

Ilkow, C. S., M. Marguerie, C. Batenchuk, J. Mayer, D. Ben Neriah, S. Cousineau, T. Falls, V. A. Jennings, M. Boileau, D. Bellamy, D. Bastin, C. T. de Souza, A. Alkayyal, J. Zhang, F. Le Boeuf, R. Arulanandam, L. Stubbert, P. Sampath, S. H. Thorne, P. Paramanthan, A. Chatterjee, R. M. Strieter, M. Burdick, C. L. Addison, D. F. Stojdl, H. L. Atkins, R. C. Auer, J. S. Diallo, B. D. Lichty and J. C. Bell (2015). "Reciprocal cellular cross-talk within the tumor microenvironment promotes oncolytic virus activity." Nat Med 21(5): 530-536.

Jhawar, S. R., A. Thandoni, P. K. Bommareddy, S. Hassan, F. J. Kohlhapp, S. Goyal, J. M. Schenkel, A. W. Silk and A. Zloza (2017). "Oncolytic Viruses-Natural and Genetically Engineered Cancer Immunotherapies." Front Oncol 7: 202.

Kamran, M. Z., P. Patil and R. P. Gude (2013). "Role of STAT3 in cancer metastasis and translational advances." Biomed Res Int 2013: 421821.

Kaufman, H. L., F. J. Kohlhapp and A. Zloza (2015). "Oncolytic viruses: a new class of immunotherapy drugs." Nat Rev Drug Discov 14(9): 642-662.

Khan, S., M. Jaggi and S. C. Chauhan (2015). "Revisiting stroma in pancreatic cancer." Oncoscience 2(10): 819-820.

Kleeff, J., M. Korc, M. Apte, C. La Vecchia, C. D. Johnson, A. V. Biankin, R. E. Neale, M. Tempero, D. A. Tuveson, R. H. Hruban and J. P. Neoptolemos (2016). "Pancreatic cancer." Nat Rev Dis Primers 2: 16022.

Klein, A. P. (2012). "Genetic susceptibility to pancreatic cancer." Mol Carcinog 51(1): 14-24.

Kohlhapp, F. J., A. Zloza and H. L. Kaufman (2015). "Talimogene laherparepvec (T-VEC) as cancer immunotherapy." Drugs Today (Barc) 51(9): 549-558.

Kota, J., J. Hancock, J. Kwon and M. Korc (2017). "Pancreatic cancer: Stroma and its current and emerging targeted therapies." Cancer Lett 391: 38-49.

Marozin, S., J. Altomonte, K. A. Munoz-Alvarez, A. Rizzani, E. N. De Toni, W. E. Thasler, R. M. Schmid and O. Ebert (2015). "STAT3 inhibition reduces toxicity of oncolytic VSV and provides a potentially synergistic combination therapy for hepatocellular carcinoma." Cancer Gene Ther 22(6): 317-325.

Mei, L., W. Du and W. W. Ma (2016). "Targeting stromal microenvironment in pancreatic ductal adenocarcinoma: controversies and promises." J Gastrointest Oncol 7(3): 487-494.

Nagathihalli, N. S., J. A. Castellanos, M. N. VanSaun, X. Dai, M. Ambrose, Q. Guo, Y. Xiong and N. B. Merchant (2016). "Pancreatic stellate cell secreted IL-6 stimulates STAT3 dependent invasiveness of pancreatic intraepithelial neoplasia and cancer cells." Oncotarget 7(40): 65982-65992.

Niu, G., K. L. Wright, Y. Ma, G. M. Wright, M. Huang, R. Irby, J. Briggs, J. Karras, W. D. Cress, D. Pardoll, R. Jove, J. Chen and H. Yu (2005). "Role of Stat3 in regulating p53 expression and function." Mol Cell Biol 25(17): 7432-7440.

Park, G. T. and K. C. Choi (2016). "Advanced new strategies for metastatic cancer treatment by therapeutic stem cells and oncolytic virotherapy." Oncotarget 7(36): 58684-58695.

Pinho, A. V., I. Rooman, M. Reichert, N. De Medts, L. Bouwens, A. K. Rustgi and F. X. Real (2011). "Adult pancreatic acinar cells dedifferentiate to an embryonic progenitor phenotype with concomitant activation of a senescence programme that is present in chronic pancreatitis." Gut 60(7): 958-966.

Reichert, M., S. Takano, S. Heeg, B. Bakir, G. P. Botta and A. K. Rustgi (2013). "Isolation, culture and genetic manipulation of mouse pancreatic ductal cells." Nat Protoc 8(7): 1354-1365.

Rossi, M. L., A. A. Rehman and C. S. Gondi (2014). "Therapeutic options for the management of pancreatic cancer." World J Gastroenterol 20(32): 11142-11159.

Schober, M., R. Jesenofsky, R. Faissner, C. Weidenauer, W. Hagmann, P. Michl, R. L. Heuchel, S. L. Haas and J. M. Lohr (2014). "Desmoplasia and chemoresistance in pancreatic cancer." Cancers (Basel) 6(4): 2137-2154.

Seymour, L. W. and K. D. Fisher (2016). "Oncolytic viruses: finally delivering." Br J Cancer 114(4): 357-361.

Sharma NK, S. S., Srivastava RK (2014). "STAT3 as an emerging molecular target in pancreatic cancer." Gastrointestinal Cancer; Targets and Therapy 2014(1).

Shen, R., Q. Wang, S. Cheng, T. Liu, H. Jiang, J. Zhu, Y. Wu and L. Wang (2013). "The biological features of PanIN initiated from oncogenic Kras mutation in genetically engineered mouse models." Cancer Lett 339(1): 135-143.

Siddiquee, K., S. Zhang, W. C. Guida, M. A. Blaskovich, B. Greedy, H. R. Lawrence, M. L. Yip, R. Jove, M. M. McLaughlin, N. J. Lawrence, S. M. Sebti and J. Turkson (2007). "Selective chemical probe inhibitor of Stat3, identified through structure-based virtual screening, induces antitumor activity." Proc Natl Acad Sci U S A 104(18): 7391-7396.

Simovic, B., S. R. Walsh and Y. Wan (2015). "Mechanistic insights into the oncolytic activity of vesicular stomatitis virus in cancer immunotherapy." Oncolytic Virother 4: 157-167.

Singh, P. K., J. Doley, G. R. Kumar, A. P. Sahoo and A. K. Tiwari (2012). "Oncolytic viruses & their specific targeting to tumour cells." Indian J Med Res 136(4): 571-584.

Storz, P. (2017). "Acinar cell plasticity and development of pancreatic ductal adenocarcinoma." Nat Rev Gastroenterol Hepatol 14(5): 296-304.

Teng, Y., J. L. Ross and J. K. Cowell (2014). "The involvement of JAK-STAT3 in cell motility, invasion, and metastasis." JAKSTAT 3(1): e28086.

von Ahrens, D., T. D. Bhagat, D. Nagrath, A. Maitra and A. Verma (2017). "The role of stromal cancer-associated fibroblasts in pancreatic cancer." J Hematol Oncol 10(1): 76.

Wojton, J. and B. Kaur (2010). "Impact of tumor microenvironment on oncolytic viral therapy." Cytokine Growth Factor Rev 21(2-3): 127-134.

Xie, D. and K. Xie (2015). "Pancreatic cancer stromal biology and therapy." Genes Dis 2(2): 133-143.

Xiong, A., Z. Yang, Y. Shen, J. Zhou and Q. Shen (2014). "Transcription Factor STAT3 as a Novel Molecular Target for Cancer Prevention." Cancers (Basel) 6(2): 926-957.

Xu, M., B. P. Zhou, M. Tao, J. Liu and W. Li (2016). "The Role of Stromal Components in Pancreatic Cancer Progression." Anticancer Agents Med Chem 16(9): 1117-1124.

Xu, Z., S. P. Pothula, J. S. Wilson and M. V. Apte (2014). "Pancreatic cancer and its stroma: a conspiracy theory." World J Gastroenterol 20(32): 11216-11229.

Yaqing Zhang, W. Y., Meredith A. Collins, Filip Bednar, Sabita Rakshit, Bruce R. and B. Z. S. Zetter, Ivy Chung, Andrew D. Rhim and Marina Pasca di Magliano (2013). "Interleukin 6 Is Required for Pancreatic Cancer Progression by Promoting MAPK Signaling Activation and Oxidative Stress Resistance." Cancer Res 73(20).

Yu, W. and H. Fang (2007). "Clinical trials with oncolytic adenovirus in China." Curr Cancer Drug Targets 7(2): 141-148.

Yue, P. and J. Turkson (2009). "Targeting STAT3 in cancer: how successful are we?" Expert Opin Investig Drugs 18(1): 45-56.

Zanoni, M., F. Piccinini, C. Arienti, A. Zamagni, S. Santi, R. Polico, A. Bevilacqua and A. Tesei (2016). "3D tumor spheroid models for in vitro therapeutic screening: a systematic approach to enhance the biological relevance of data obtained." Sci Rep 6: 19103.

Zervox, E. E., M. G. Franz, K. F. Salhab, A. E. Shafii, J. Menendez, W. R. Gower and A. S. Rosemurgy (2000). "Matrix metalloproteinase inhibition improves survival in an orthotopic model of human pancreatic cancer." J Gastrointest Surg 4(6): 614-619.

Zhu, Q., X. Zhang, L. Zhang, W. Li, H. Wu, X. Yuan, F. Mao, M. Wang, W. Zhu, H. Qian and W. Xu (2014). "The IL-6-STAT3 axis mediates a reciprocal crosstalk between cancer-derived mesenchymal stem cells and neutrophils to synergistically prompt gastric cancer progression." Cell Death Dis 5: e1295.